TLC
ADVENTURES
for your mind
INTIMATE
UNIVERSE
The Human Body

INTIMATE UNIVERSE

The Human Body

Anthony Smith

DISCOVERY BOOKS
RANDOM HOUSE, INC.

To Richard Dale
and his television team

This book is published to accompany the television series
Intimate Universe: The Human Body produced for TLC (The Learning Channel)
and BBC Television by the BBC Science Unit.

Executive Producers: Alan Bookbinder and Lorraine Heggessey
Series Producer: Richard Dale

This work was originally published in Great Britain by BBC Books in 1998.

ISBN 0-679-46251-1

Printed in Great Britain

98765432

First US Edition

Commissioning Editor: Sheila Ableman
General Editor: Jo Whelan
Picture Researchers: Susannah Parker and Deirdre O'Day
Art Director: Linda Blakemore
Designer: Rachel Hardman Carter

Set in Janson by BBC Books
Printed and bound in Great Britain by Butler & Tanner Limited, Frome and London
Colour separations by Radstock Reproductions Limited, Midsomer Norton
Jacket printed by Lawrence Allen Limited, Weston-super-Mare

Contents

Chapter 1 **Introduction 7**

Chapter 2 **The Beginning 23**

Chapter 3 **Childhood 53**

Chapter 4 **Puberty 85**

Chapter 5 **Adulthood 119**

Chapter 6 **Ageing 155**

Chapter 7 **Death 185**

Bibliography 211

Picture credits 212

Index 213

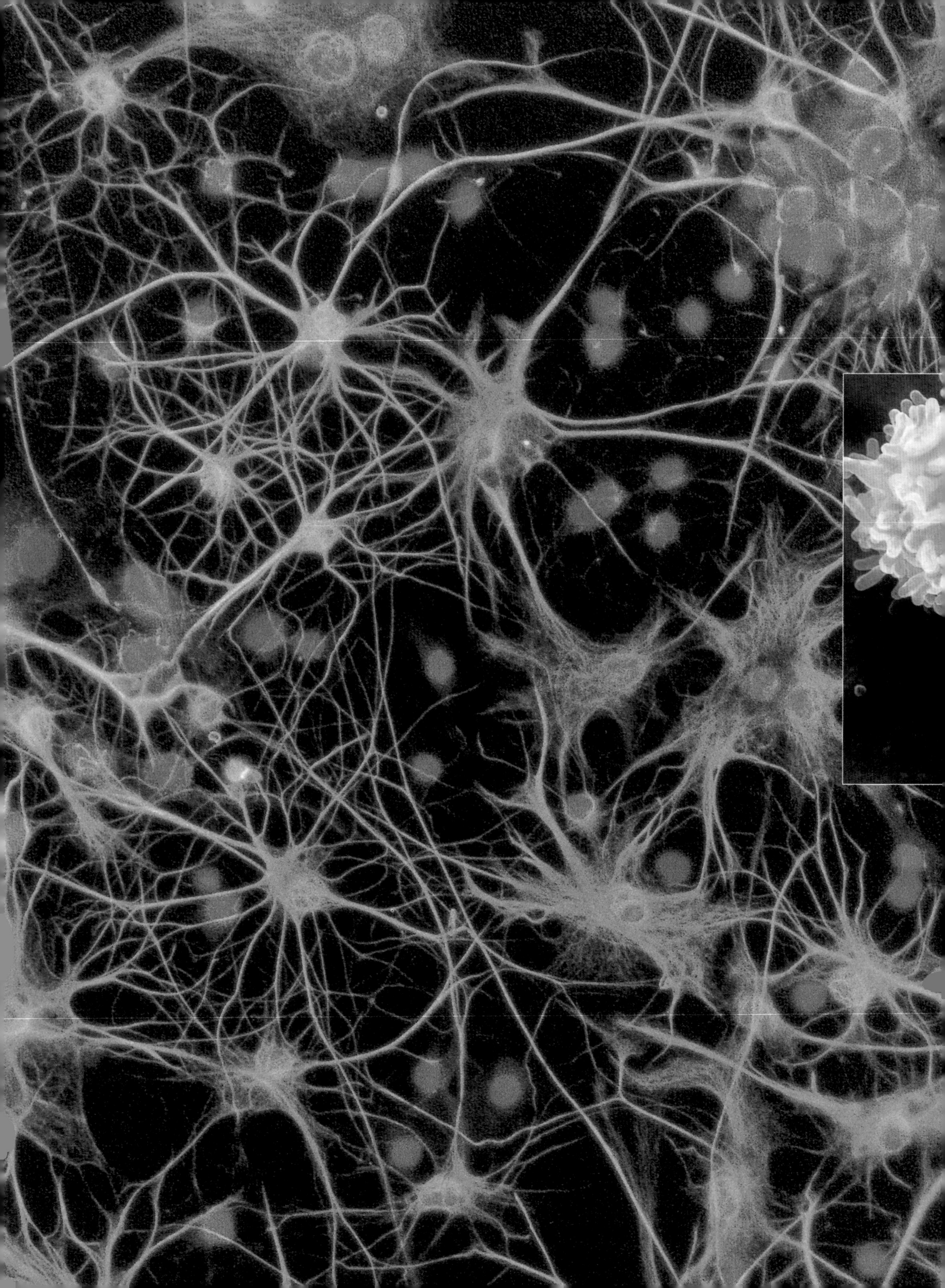

INTRODUCTION

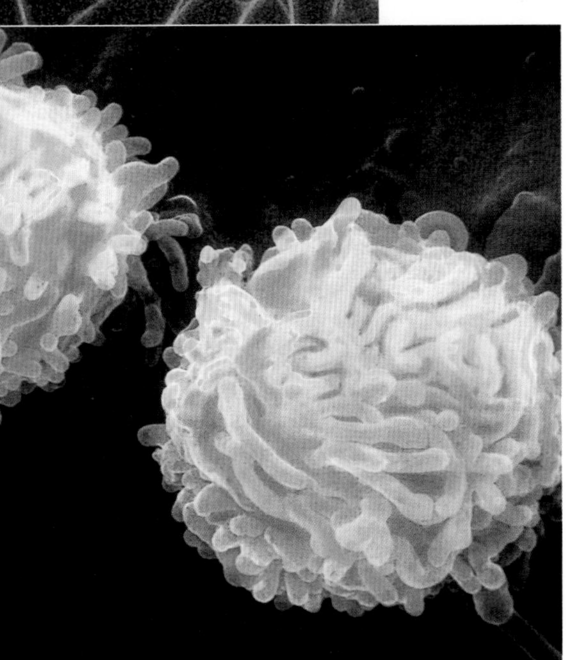

Another baby is born, 3 kilograms of potential. Its body is a collection of assorted chemicals: around 10% protein, 10% fat, 1% sugar and 75% water. Their assembly has been orchestrated by the 100,000 genes the baby has inherited from its parents. These have determined what it has become, and they will continue to organise its body throughout life.

And what a life it will be. A new individual born in the developed world can expect to live for about 75 years, 3 years less if it's a male and 3 years more if a female. During that time this helpless, dependent bundle – let's call it a 'she' – will transform herself many times over. From infant to child to adolescent to adult, and eventually through to ageing and death, she will face and adapt to a continuous series of challenges.

Her body will grow until it contains ten thousand billion cells, nearly all of which will undergo a constant cycle of death and renewal. During her life she will spend five solid years eating and drinking to supply the amazing machine in which she lives with fuel. Her first words will be uttered at about 12 months after birth, and by the age of six she will have mastered the essentials of language. So important will this skill prove that up to ten years of her life will be spent talking. She will be able to put names to around two thousand faces, but will only consider about 150 of these as friends. She will have sex over three thousand times, with perhaps eight different partners, and will have two children and four grand-children. Her feet will walk fourteen thousand miles

Left Micrographs of brain cells (*main picture*) and white blood cells (*inset*). These and many of the other pictures in this book have been taken using sophisticated electronic imaging technology. The colours in these images are generated by computer for the purposes of clarity and visual interest, and do not represent the true colours found in the body.

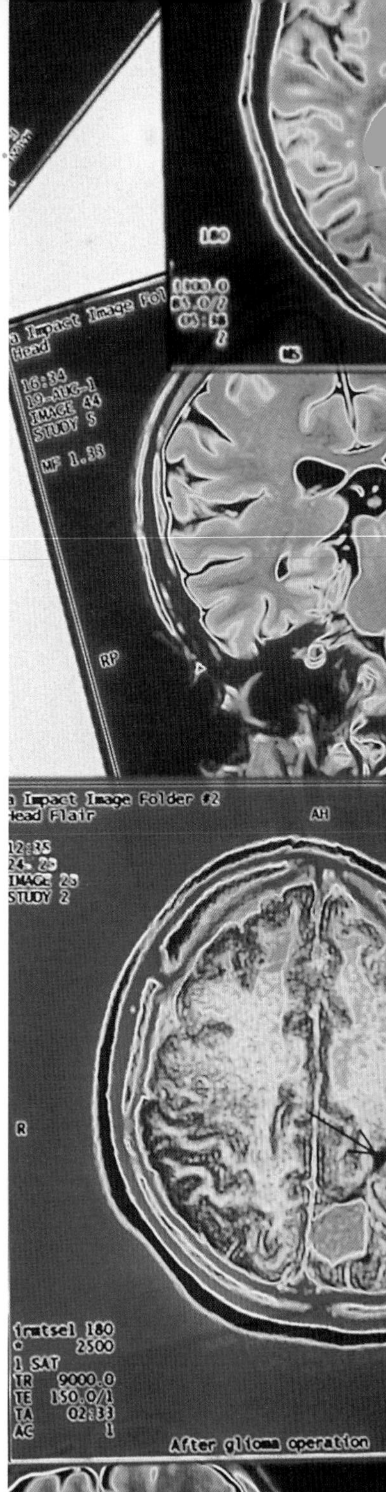

and she will spend a year travelling in a car. Paid work will account for another nine years at most. To recover from all this, she will spend 22 years sleeping.

Despite all our activities and achievements, nearly everything our bodies do is a secret from us. In reality, what we see as the normal monotony of living, in which nothing much happens for most of the time, is the result of an intricate balancing act of which we are almost completely ignorant. Even when we think we are relaxing, say sitting down to read a book, our body is vigilant and busy. Nerve endings in the skin send signals to the brain to ensure we are sitting comfortably in the chair, and we absent-mindedly tug at the cushions until the messages say our position is satisfactory. Meanwhile another part of the brain is processing input from our eyes, converting the little arrangements of ink on the white page into words. Somewhere else again the words are given meanings, and the ideas put down by the book are passed on.

Perhaps we had a meal before settling down. If so, our stomach is busy with the first stages of digestion, breaking the food down with enzymes and acids before releasing it in small bursts to the small intestine. Blood flow in our muscles is diverted towards the intestines so that the nutrients can be absorbed. The broken down components of our food are taken away and distributed to individual cells, and the body is kept in fuel and raw materials for another day.

Sitting in our chair we are performing no muscular work beyond holding ourselves upright, so our oxygen requirements

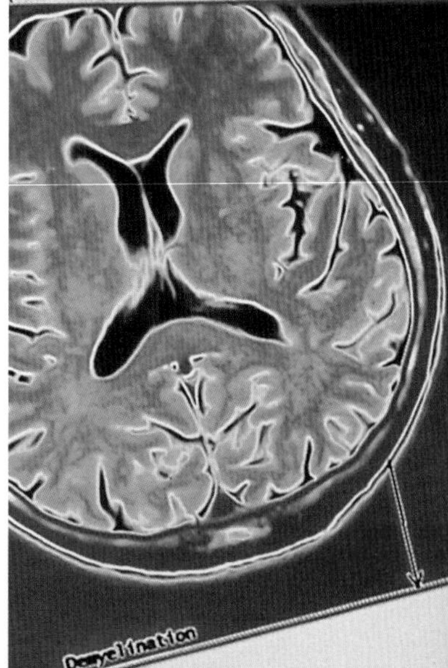

Right MRI (magnetic resonance imaging) scans of the human body. These show 'slices' through the body, created from radio signals in a powerful magnetic field. This and other non-invasive techniques (such as CAT and ultrasound) permit an investigation of tissues unthinkable even a few years ago. They can lead to better diagnosis and treatment planning for a wide range of diseases.

Impact Image Folder #1
Impact Image Folder #2
MAGNETOM IMPACT
H-SP-CR
3800.0
120.0/1
180

are at base level, and the brain stem adjusts the heart rate and breathing accordingly. But it cannot last. That cup of coffee we had half an hour ago has already been absorbed from the intestine and has diluted our blood. The excess water has been filtered out by our kidneys and has accumulated drop by drop in the bladder, causing it to stretch. Stretch-sensitive nerves in the bladder begin signalling to the brain, and we realise we need to relieve ourselves. As we get out of the chair, the brain reacts to signals from the ear and controls the tension of our muscles so that we don't fall over. The circulation adjusts to the fact that we are now upright – though getting up too quickly might have made us dizzy for a second before the system caught up. Climbing the stairs to the bathroom, the heart's output is increased to supply the leg muscles with more oxygen, which is obtained by breathing a little more deeply. We don't notice any of these adjustments.

Later in the evening powerful chemical messengers flood the brain, and we begin to feel distinctly sleepy. But the body never rests, even during sleep. Its mission at all times is to maintain its internal environment within the narrow boundaries compatible with healthy existence. Temperature must be regulated, acidity and water balance controlled, invading bacteria dealt with. Worn-out tissues must be replaced and wounds healed. Cells must be fed and their waste products removed. A thousand times a second, chemical battles are waged on behalf of our continued existence, but we remain in blissful ignorance.

The body's automatic workings are controlled by specific areas of the brain, leaving other areas free for control of voluntary activity, interpretation of information received (sensory processing) and abstract thought and emotions. These 'higher' functions are carried out in the cerebral hemispheres, the ballooning lobes at the top and front of the brain. Animals such as rats or dogs are just as good or better than humans at the automatic functions, but their cerebral hemispheres are puny. What made humans as clever as they are was the huge expansion of the hemispheres over evolutionary time, resulting in the most capable brain there has ever been.

Animals' bodies are just as amazing as our own, and sometimes more so. But we are unique because we can wonder at the way we are made. We are aware of our ignorance and seek to unravel the body's mysteries. Thanks to modern science, we are now able to see inside ourselves in a way no generation in history has ever done before. New imaging techniques (see pages 8–9) show us the body and even the brain in action, but they only heighten the sense of wonder and of ignorance. 'Before I came here,' said the brilliant physicist Enrico Fermi, 'I did not understand your subject. I still don't understand it, but on a higher level.'

One area where our understanding has grown exponentially in the last 20 years or so is genetics. Since the 1960s it has been known that the genetic information that shapes us is carried on the molecule called deoxyribonucleic acid, abbreviated to DNA (see page 154). DNA looks like a ladder that has been twisted into a spiral. The rungs of the ladder can be made from any of four different chemicals, and it is the sequence of these chemicals that encodes the genes. Humans have about 100,000 genes, each consisting of many thousands of chemical rungs. The whole library is carried in every cell, split into 46 units called chromosomes (the chromosomes of a male are shown on pages 12 and 13).

Despite carrying our entire genetic blueprint, the DNA in each cell weighs no more than 6 millionths of a millionth of a gram. This figure is not much more comprehensible when multiplied by the number of people on the planet. Each of the 5.8 billion humans now alive started off as a single set of DNA in a fertilised egg. Therefore, the genetic material that initiated every one of them only weighed, in total, 1/30th of a gram.

It is hard to comprehend that this speck of DNA was able to start the process which led to our fingernails and toenails, our eyebrows and eyelashes, and to our whole array of working organs, including the 50 billion cells of the brain. Back-tracking even further, we remember that our genetic inheritance came from our mother and father. Their germplasm, the primordial cell material which divides to form sperm and egg, was laid down during the first month of their own development, long before their births. And still the line goes back. The hominid individuals whose fossils we sometimes unearth gave rise to modern humans. Similarly their ancient ape-human forebears came from primate ancestors, which in turn arose from some other mammal much earlier.

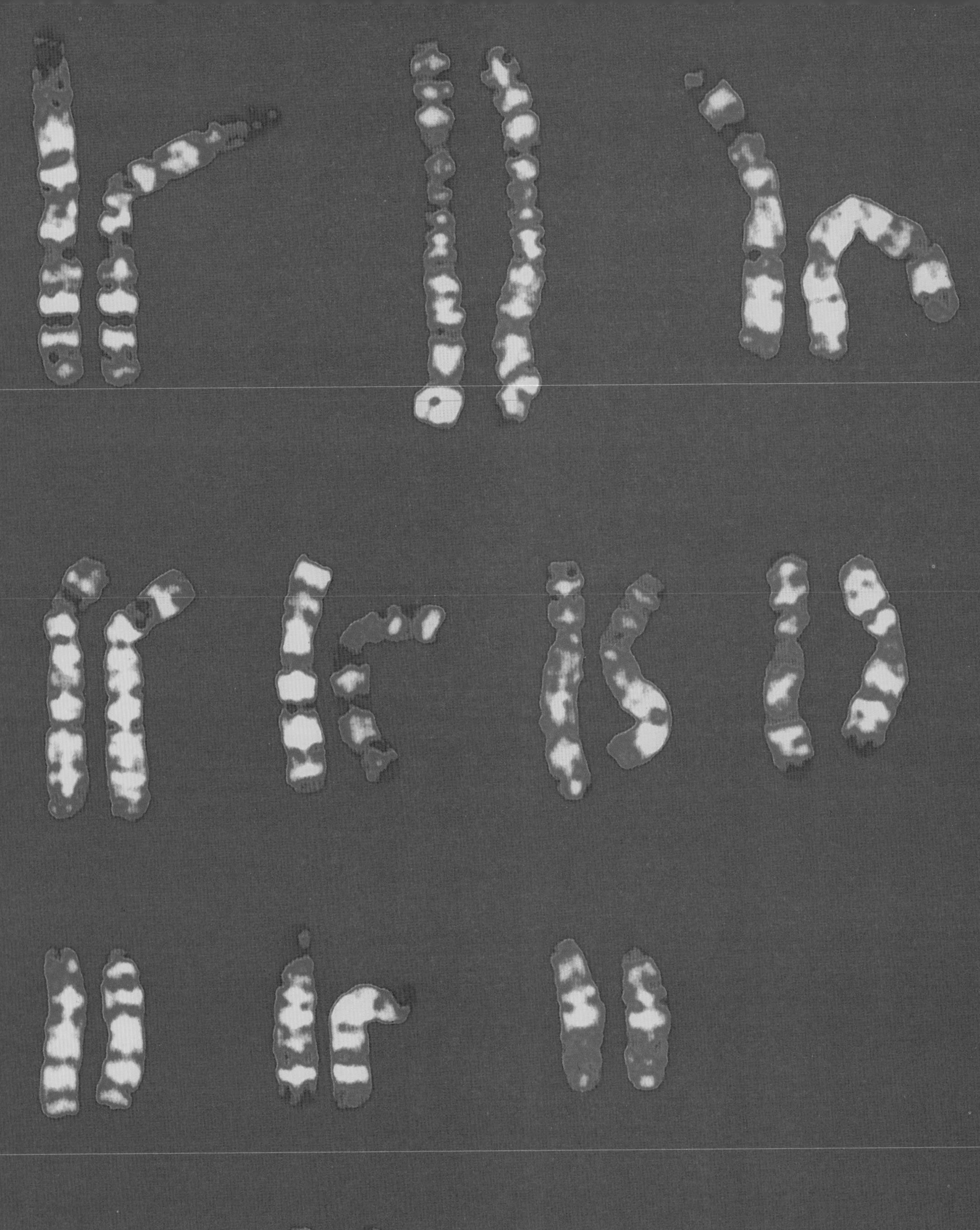
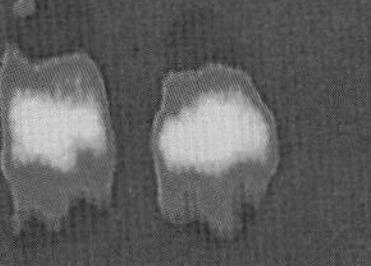
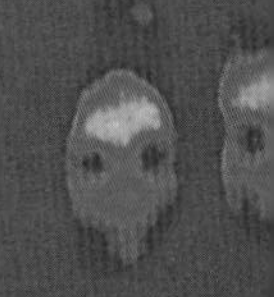

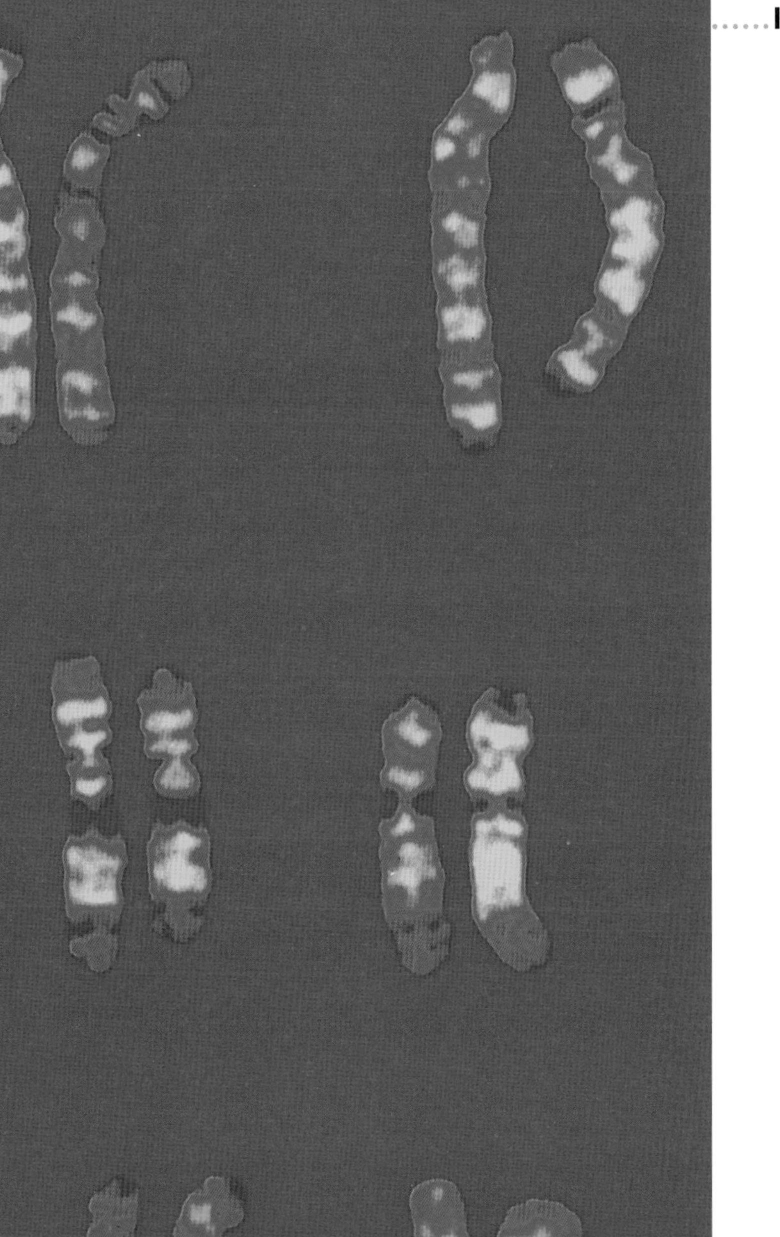

Left The 23 pairs of human chromosomes, here arranged in size order, are numbered from the largest (number 1, top left) to the smallest (number 22) and store the genetic information specific to each individual. Each member of a pair is similar in length and appearance except for the sex chromosomes of the male (number 23, bottom right) which are long and short (known as X and Y). The female sex chromosomes are both long – women possess two X chromosomes. These sex chromosomes determine a person's gender.

The story of evolution, from bacteria to amoebae to jellyfish, from fish to amphibia to reptile to mammal, is both stimulating and emotive. Are humans no more than the product of an entirely blind sequence of events? Is there any critical difference between us and the other animals? Our consciousness, and the possession of a brain that encourages us to ask such questions, can make us believe that there is such a difference. This is a matter for individuals and for personal belief. On a purely biological level, the evidence of evolution suggests our origins were from simpler and humbler forms of life.

In Mud Creek, part of America's Yellowstone National Park, the ground bubbles and hisses. It seems alien to life. To walk along the conveniently placed duck-boards is an unnerving experience – a misplaced foot could plunge painfully into the hot liquid. There are no trees, no bushes and no grasses growing in this environment. There are not even insects, let alone fellow mammals; but, strangely, there is life.

Much of the planet earth may have resembled Yellowstone's Mud Creek three billion years ago. The land then was hotter than now, but simple life forms – such as algae – were able to survive in such conditions. They still do, and the hot Yellowstone mud is thick with them. They actually thrive on the sulphur and carbon dioxide that belch from the earth, and would die if transplanted somewhere else. Any human observing them should remember that these algae, resembling little more than slime, were once the most advanced forms of life on the planet.

It is arguable that these heat- and sulphur-loving algae have been extremely successful. They have lived, more or less unchanged, for three billion years. By comparison, *Homo sapiens* has existed for only a minute fraction of that time. On the other hand, these algae have been, to put it bluntly, stick-in-the-muds. They have not altered. They can only live in the environment they first encountered. If that changes, if mud creeks disappear from the earth, they will perish. The process of evolution encourages the ability to change, to be adaptable, and to survive in altered circumstances. By that yardstick the Yellowstone algae are failures. Only a few parts of the world have stayed as boiling, sulphurous mud, and only there can they exist.

For evolution to have transformed such simple creatures into such complex ones (human beings, for example) can seem impossible. The adaptations necessary seem too numerous, too complicated, the leap too great. The idea only becomes plausible when we grasp the huge lengths of time involved, and the gentle, step-by-step nature of the change. Three billion years is unimaginably long to most humans (who, by living just 27,000 days or so, do not even last for 3 billion *seconds*). The many changes which took place in transforming, say, a family of amphibians into reptiles, were each of themselves tiny. But over the aeons of time and via countless generations, these changes added up to major transformations.

Evolutionary changes are alterations of something that already exists. Thus the fish's swim-bladder, a balloon-like object for maintaining buoyancy, became lungs for the later air-breathers, for whom buoyancy was no longer an issue but obtaining oxygen from gaseous air was crucial. So too for the limbs of four-footed animals. The original pentadactyl (five-digit) form evolved into the single digit of horses, the double digit of cattle and antelopes, the wings and feet of birds and bats, and the clever, manipulative hands of humans (along with our relatively clumsy feet). As somebody once put it, nature started with a bicycle and produced a sewing machine.

An interesting example of evolutionary adaptation comes from the three bones of the human middle ear, the smallest bones we possess. Known as the malleus, incus and stapes after their slight resemblances to a hammer, an anvil and a stirrup, they sit in the middle ear cavity, an air-filled hole about 8 millimetres wide and 4 millimetres deep. Their task is to pass on sound vibrations from the ear-drum to the nerves of the inner ear, which in turn pass the details to the brain.

So far so good, but where did these three crucial bones come from? A reptile has only one auditory bone in its ear, but its jaw contains extra bones compared with mammals. These, in part, were used during the long passage of evolution in the creation of the mammalian ear. The reptilian jaw bones were themselves amendments of earlier structures that are used to support the gills in fish. When some fish abandoned the water they no longer needed gills, and the structures became available for other purposes.

In a five-week human foetus evidence for this ancient story is clearly visible. Near the head there are six cartilages on either side, neatly and symmetrically arranged like the gill supports of a fish. Beside them are blood vessels and muscles, lined up as though destined to join the gill system. When the foetus reaches 20 weeks this earlier layout has been totally transformed. All six cartilages have acquired new roles. The first, nearest the head, becomes the incus, malleus, and some pieces of jaw. The second becomes the stapes and the hyoid bone (which provides tongue support). The third also forms the hyoid, and the fourth is used to form much of the larynx. The sixth also contributes to the larynx, and the fifth disappears. The orderly grouping of related blood vessels is gone too, remoulded to fulfil completely different purposes.

As biologists like to say, 'ontogeny recapitulates phylogeny'. This means that the development of an individual reflects its evolutionary past to some degree. A very early human embryo is not identical to a fish or a reptile embryo, but there is a degree of resemblance. Only as the genetic program runs on does each species begin to develop the characteristics that make it unique.

Given the staggering changes that evolution has produced, we might expect to see evidence of the process in our own species on a timescale we can comprehend. Instead, evidence indicates that human beings have not changed significantly over such time scales. Cave-dwellers lived by hunting animals and gathering fruits, and knew nothing of writing, house-building or agriculture, or of the planet on which they lived. Their cave wall pictures were painted thousands of years

Left Within the human middle ear, lie three bones, the malleus, the incus and the stapes responsible for transmitting sound waves to the nerves of the inner ear. They are thought to have evolved from the jaw bones of our amphibian ancestors.

before Stonehenge was erected, before the Egyptian pyramids, before all of history.

Yet, if the paintings have been accurately dated at 20,000 years before present, only 800 generations separate them from us. Their bodies were probably not quite identical to ours, but virtually so. Their brains were certainly the same size – human cerebral capacity has not changed for 100,000 years. If, magically, one of their newborn babies could be transported to modern times he or she would not be some alien individual, but would learn to read and write, speak the local language, and become just another ordinary human being. In the same way, a modern baby transported to the cave would grow up as an ordinary hunter-gatherer without any special abilities.

Bears also gather fruit and nuts, hunt and shelter in caves, but possess pathetic brains by comparison with the swollen hemispheres of *Homo sapiens*. Yet the 1400 cubic centimetres of modern brain, enabling humankind to build skyscrapers, create aircraft, write symphonies and explore space, was developed for individuals who did none of these things. The human brain is an enigma. It is not the largest in the animal kingdom, nor the biggest in proportion to body size, but it is far and away the cleverest.

Part of this large brain's role is to help each individual change throughout life to meet new challenges. The baby's goals are first to make sure that others meet its needs, and second to learn the essentials of life so that it can become independent. Its whole brain is geared to picking up new skills, and grows at a phenomenally rapid rate. By the age of six or so the child has mastered everything it needs for survival and is perfectly adapted to the challenge of staying alive. But something is missing. Unable to pass on its genes to the next generation, the child is an evolutionary dead end. In order for the genes to complete their biological mission, the body must change again, this time through the process of puberty. Instead of merely staying alive, the challenge of adulthood is to reproduce and successfully rear

Right The human skeleton grows throughout an individual's life adapting itself to each new challenge from learning to walk to developing the ability to reproduce.

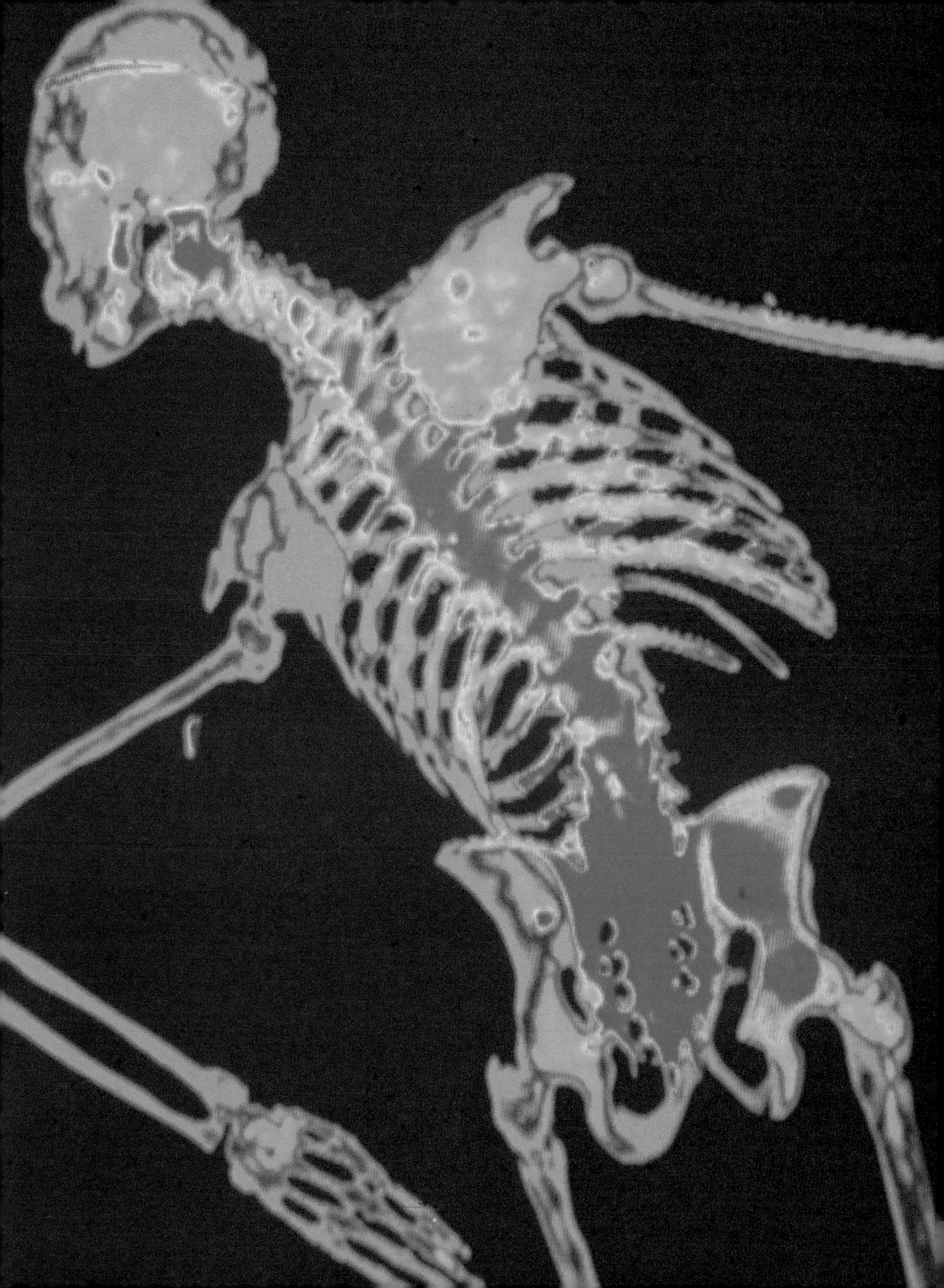

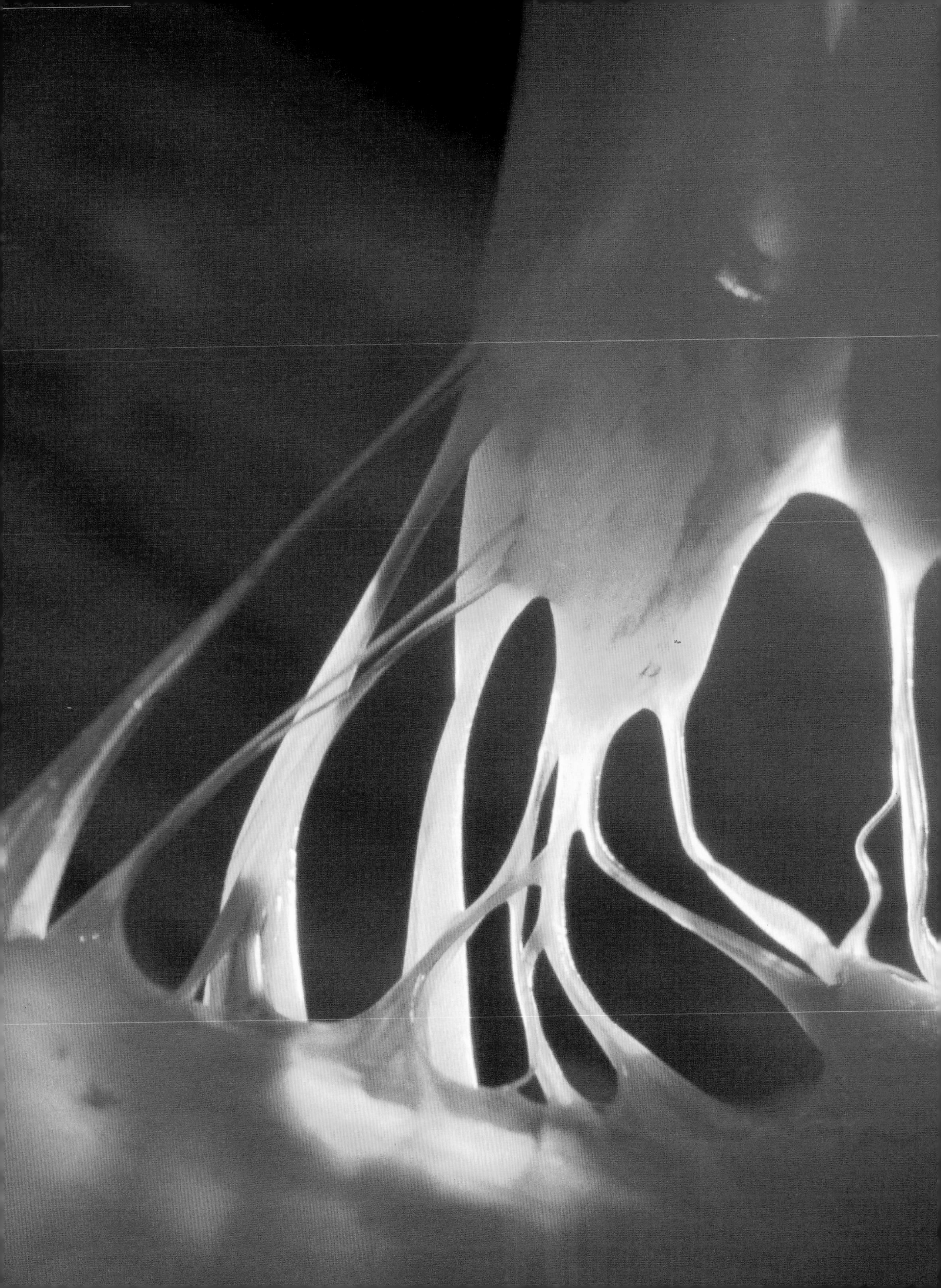

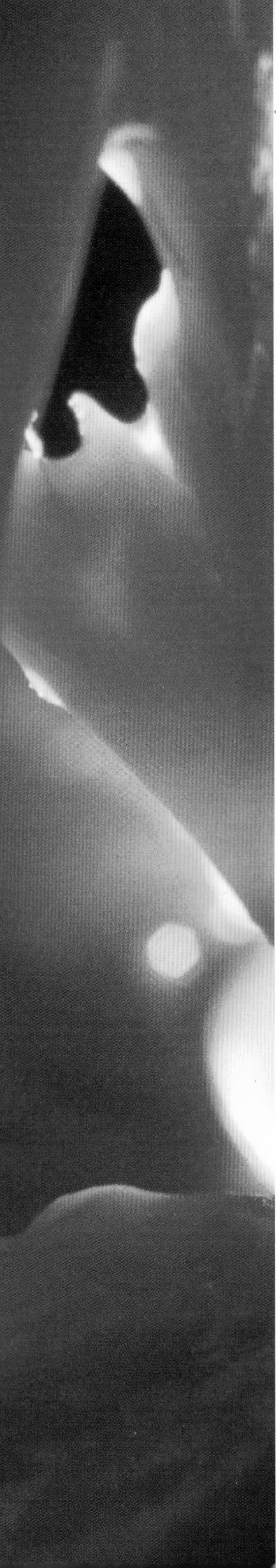

offspring. Once this is achieved the work of the genes is done. Although they keep the body going for many years afterwards, they no longer drive forward the process of change. Ageing is a gradual winding down, and the death that eventually follows is not part of a grand scheme but a consequence of deterioration.

'What a piece of work is man!' wrote Shakespeare; 'How noble in reason! How infinite in faculty!' When we think of human bodies, instead of focusing on how astonishing they are we more often think of breakdown. We are regularly annoyed by temporary infections and minor malfunctions, and occasionally devastated by major ones. This book concentrates upon the wonders of the human body. These are amazing. Or, rather, each one of us is amazing.

During our lifetimes our bodies are us. None of our achievements, ideas or emotions would be possible without them. We can carry on the business of life in almost total ignorance of how they work, but to do so is to block out a whole dimension of our existence. Young or old, male or female, black or white, we all have the unifying feature of a body.

The story of the human body has been spelled out over millions of years of evolution, and is also told by each of us in our own lifetime. We live it, and yet we do not know it as intimately as we might. For all human beings it is the greatest story ever told. And it begins with those two cells – the biggest in the human body and one of the smallest – namely an ovum and a spermatozoon, or an egg and a sperm.

Left A portion of the heart valve. The pumping of the blood around the body by the heart is an automatic working of which we are unaware for the majority of our lives.

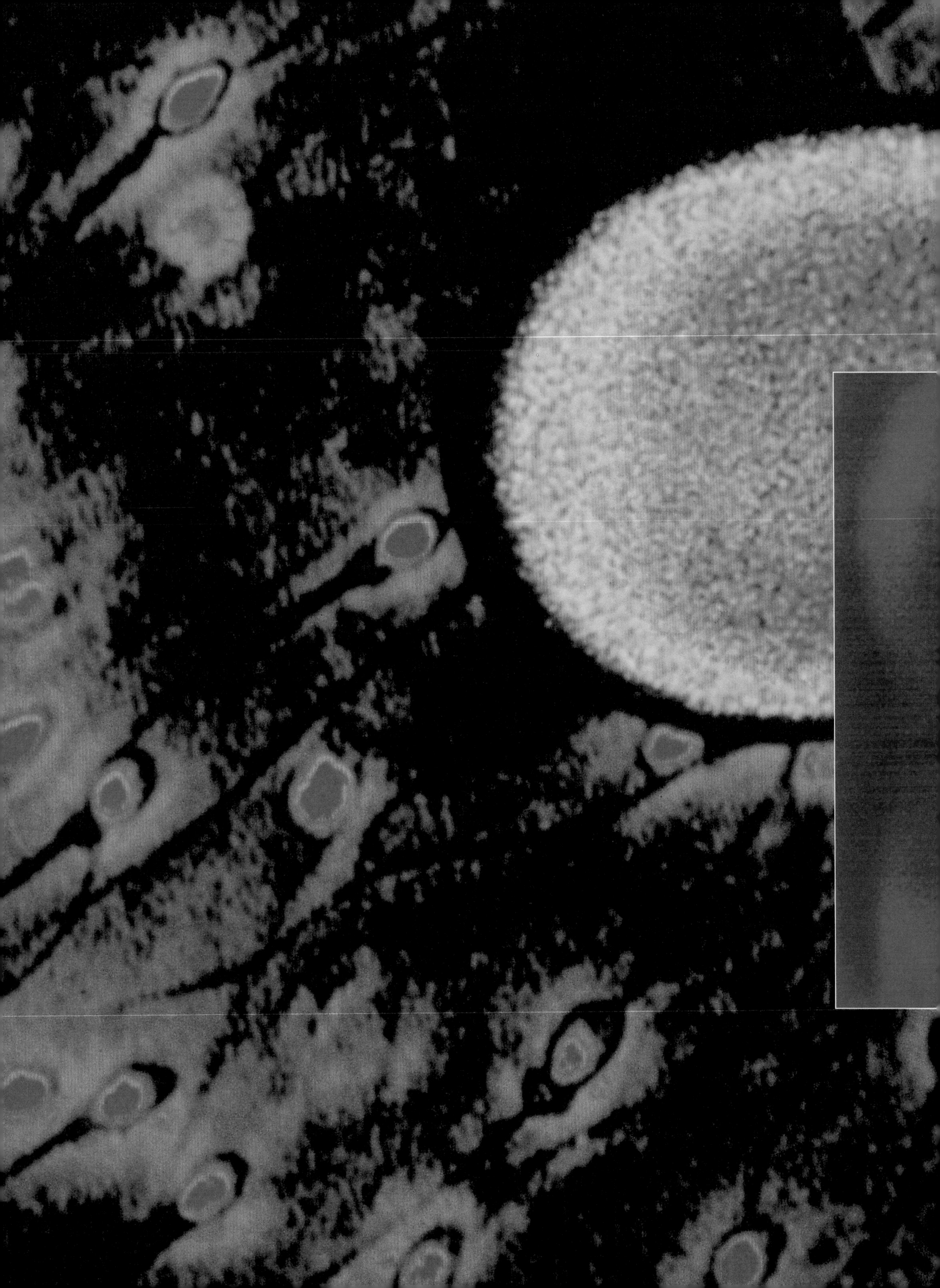

The BEGINNING

In one sense it is a love story. There can never be, within the sexual system of reproduction, a closer bonding between one individual and another. For three-quarters of a year two individuals are attached, one developing within the other. Intimate closeness continues afterwards through the baby's utter dependence on feeding and care, but the earlier union is beyond compare with any of the other partnerships either may meet in life.

The second parent of any child, the father who contributed 50% of its genetic material, may be emotionally affected, and caring, and concerned, but he is as distanced physically as is every other observer of the pregnancy. He may lavish love and attention when the child is born, but he cannot suckle the infant any more than he could give it sustenance for the earlier nine months. He is an outsider to the events he was instrumental in creating.

Yet the mother is also an outsider, despite the intimacy of the union. The child to whom she gives birth is not a piece of her, like some bud extruded from a main stem. It is a separate human being, a stranger however close, and as distinct genetically from her as are any of its siblings from each other. A mother, for example, cannot usually give transplanted material

Main picture Sperm approaching the egg. There may be 200 million sperm in an ejaculation, but there is enormous attrition as they make their way via the uterus to the Fallopian tube. Only a few hundred may reach the vicinity of the egg. Despite the difference in their sizes, both sperm and egg carry the same quantity of genetic material.

Inset Between conception and birth: a digitised image of a ten- to fifteen-week foetus.

PHILLIPPA

Phillippa first met Jeff of Bath, England in May 1989 and they married a year later. By the time their thoughts of children moved from vague wishes to more positive intent, Phillippa was in her mid-thirties. However, according to Jeff, they were 'both a little naive about the difficulty of having a child' and conceiving took longer than they thought. Phillippa was concerned. 'I felt the biological clock was ticking away. I've always wanted to have children, and it's easy to worry that you've left it too late.'

Fertility does diminish for potential mothers as the clock ticks on. Phillippa and Jeff were typical of modern times, delaying their first pregnancy. The 1990s witnessed, for the first time, more births to British women in their thirties than in their twenties.

Phillippa and Jeff did eventually conceive, and later were thrilled to receive visual confirmation of the pregnancy through an ultrasound scan. The developing foetus was then 12 weeks post-conception. They were both astonished at what they saw. For Jeff, it really brought home that Phillippa was actually pregnant. She had not been sick. There was no bump at the front. He was amazed to realise that there *was* something in there. 'It was kicking, and waving its arms – incredible'.

Towards the end of her pregnancy Phillippa said the whole experience had been 'quite wonderful', and that she had thoroughly enjoyed it. There had been no significant problems and she was very excited about impending motherhood, although admitting to occasional feelings of panic at the awesome commitment. However, as the birth grew closer she began to feel breathless, with touches of indigestion and heartburn as the baby pressed against her internal organs. Instead of sleeping on her back she had to lie sideways.

Then she went into labour. At first she was not sure if the waters were breaking or if the baby's head was pressing on her bladder. Contractions settled the matter. Some were 'not too bad', while others made her want to 'screech, scream and clench her teeth'. Much pushing later the baby was born, a girl; the parents were delighted.

to her progeny. Nor can they donate any of their substance, such as skin, to her. All such offerings will be rejected and destroyed, as are all transplants detected as foreign by the recipient. And yet, a mother's offspring are entirely created within her.

Half a century ago, during the early days of understanding the rejection mechanism, it was argued, tongue in cheek, that human babies are impossible. Is the foetus anatomically isolated? Is the mother's placenta, or even her whole body, immunologically inert, incapable of rejecting foreign tissue? With the answer to each question an emphatic 'no', the puzzle remained. Today, with the rejection process better understood, the central conundrum is still in place. We still do not know how a baby can be created without the mother's body rejecting it.

The fact of a baby being initiated, nourished and then delivered can seem more remarkable to many a parent than the fact of their own existence. All of us eat, digest, breathe, walk, hear, see, smell, feel, speak, excrete, think, reason and anticipate as we go about our normal tasks, and pay little attention to what our bodies are doing. Pregnancy can make us think, more than ever before, about the astonishing processes that we normally take for granted.

Pregnancy also has its disagreeable side. There may be nausea and physical discomfort, as if some illness is in charge of the women's body. There can be shifts in mood as hormones go into overdrive. The mother's body gains an average of 12.2 kilograms or 27 lb in weight, protruding grossly in a way few other animals seem to match. The whole business, this nurturing of a parasite, is less than immediately appealing to some. At the end of it all, after 38 long weeks of increasing physical distortion from normality, there is the disturbing prospect of the birth, usually painful and unpredictable in its timing. 'Let's face it,' said Shulamith Firestone, US feminist, 'pregnancy is barbaric.'

Pregnancy used to be a closed book, save for each mother's signs, symptoms and perceptions. She knew she might be pregnant when an expected menstrual flow did not occur (though it can fail for other reasons). She most probably experienced nausea, (two-thirds of women do). Her breasts became tender and then grew, with the areolae darkening. Later, as the final proof, she could feel movement, 'the quickening', starting around the eighteenth week. Weight gain and size increase would be additional confirmation (and hard to conceal). As for the happenings within, they were a mystery. Only on the final day, the 266th or so since it all began, would she learn some facts about the previous nine months. The finished product would then be in her arms.

When science first prised open the secret happenings of pregnancy, the early procedures smacked more of apothecaries than modern medicine. The first hormonal pregnancy test used immature female mice. Some of the woman's urine was injected into the mice and, after they were killed 100 hours later, their ovaries would show whether she was pregnant. The process was later speeded up with the use of virgin doe rabbits, whose ovaries would alter within 48 hours. A still shorter wait became possible by using female specimens of *Xenopus*, the clawed toad. A mere 15 hours after receiving a pregnant woman's urine these amphibians would abruptly spawn a huge quantity of eggs. Such procedures were only abandoned during the 1960s when chemical, rather than biological, methods of detecting pregnancy hormones were developed. Today, simple home testing kits using antibody technology can tell a woman whether or not she is pregnant right from the first day her period is missed.

The missed menstrual period, which becomes apparent two weeks after conception, is still the key alert to pregnancy. Allowing for variations in the cycle, a woman may therefore not seek confirmation of pregnancy for three weeks or more after fertilisation has taken place. By that time (and see page 42 for progress day by day) foetal eyes and ears have begun to develop. The heart is about to start its life-long labour. If the foetus were a mouse it would already have been born. As it is a human there are 35 more weeks to go (there would be 97 for an elephant). Human beings are neither outstandingly long in their gestation time nor extremely short.

Each baby develops, on average, for 38 weeks (266 days) between fertilisation and birth. Estimated birth date is calculated by adding 280 days to the start of the last menstrual period, so for the purposes of dating the pregnancy is said to last 40 weeks. Only about 5% of babies arrive on the exact 'due' date, with up to two weeks before or after being quite normal.

Fertilisation

Each new human life starts as a single egg, viable for 8–24 hours. Having been formed before its mother's birth, it waits – for 20, 30, 40 or more years before being prepared for fertilisation. After preparation the egg is then extremely short-lived, like a bomb after its fuse is lit. At ovulation it is expelled from the ovary and crosses a small gap into the waiting tentacles of the Fallopian tube (also called the oviduct or the uterine tube). It is then wafted towards the uterus, or womb, by tiny projections on the inside surface of the tube. At this stage the egg is surrounded by a cloud of cells brought with it from the ovary, poetically named the corona radiata.

While travelling along the tube the egg either does or does not encounter spermatozoa. If it does not, it quickly dies and the menstrual cycle continues towards the eventual breakdown and expulsion of the uterine lining. The box on pages 32 and 33 looks at the male's contribution to the human reproductive process, during which sperm are produced and start their journey towards the egg.

The sperm too have waited since formation, but only for days or weeks. They can then survive for one to two days in the female reproductive tract. Of perhaps 200–400 million sperm present in an ejaculation, only a few hundred thousand will make it as far as the uterus and a thousand or so reach the Fallopian tube. By the time they reach the egg in the upper part of the tube only around a hundred are left in the race.

Sperm do not become capable of fertilising an egg until they have spent some time in the woman's body. Fluids in her reproductive tract wash away the inhibitory chemicals produced by the male to suppress sperm activity. Enzymes then set to work weakening the membrane that covers each sperm's head, and the whiplash motions of the tail become more powerful. It takes around a thousand tail strokes to drive a sperm forward a centimetre. Sperm are helped on their way by contractions of the uterus, which force them upwards into the body. The contractions are stimulated by prostaglandins (hormones) contained in the semen, and also occur during the female orgasm.

At most times of the monthly cycle, sperm may not reach the uterus at all, kept out by a plug of impenetrable mucus around the cervix. Only at the time of ovulation does the mucus loosen, becoming shot through with hair-width channels that allow the sperm to pass.

Once in the vicinity of the egg, the sperm are chemically attracted to it. When they reach their destination they must first penetrate the corona radiata, before the outer membranes of their heads bind to special receptors in the

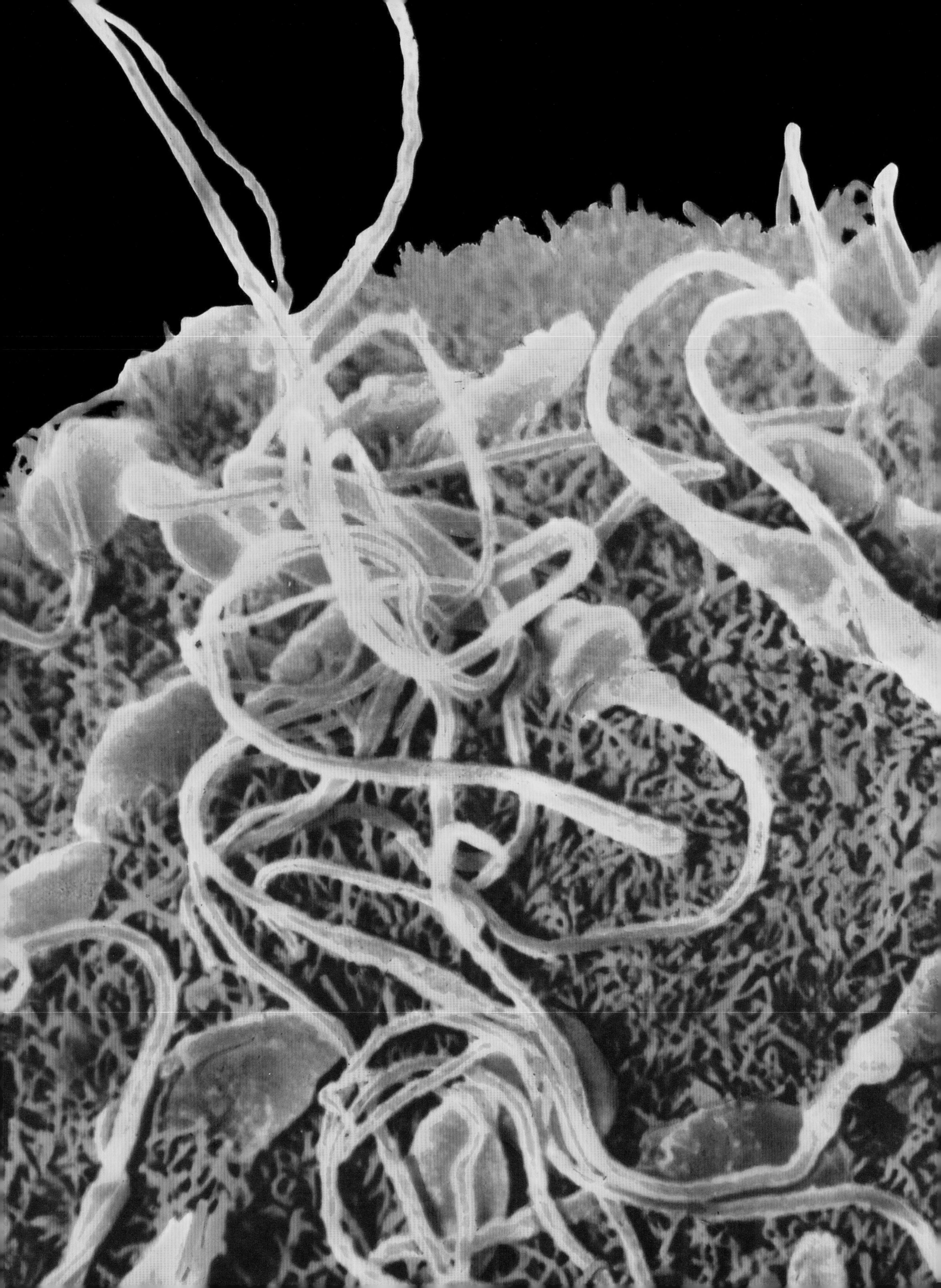

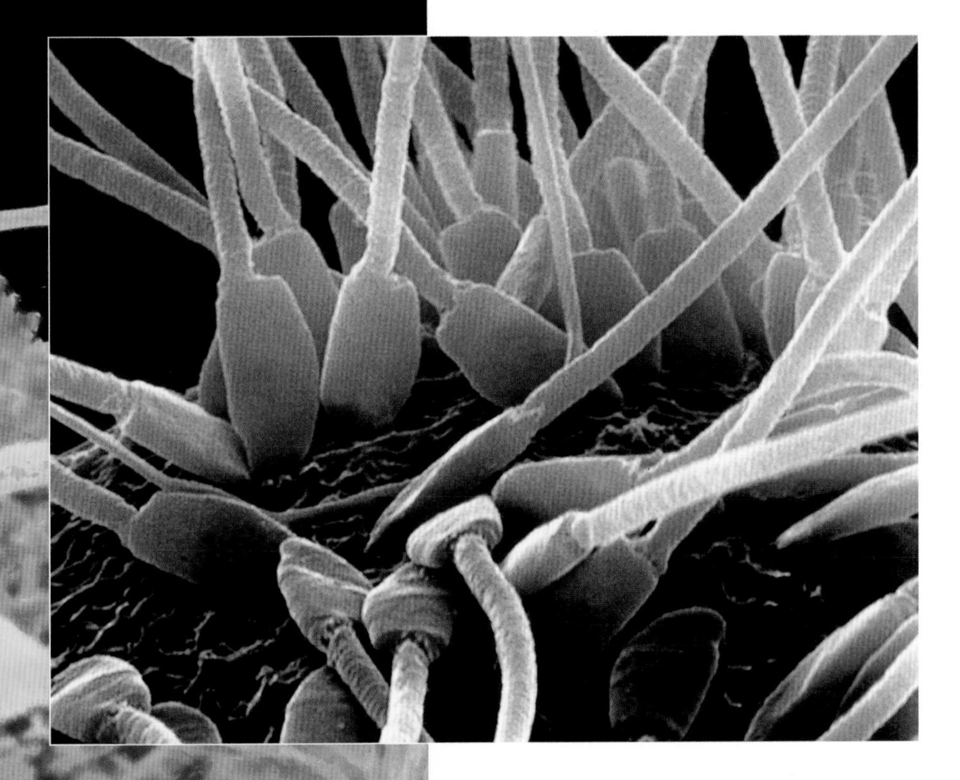

Above Towards fertilisation. The sperm are attracted to the thick surface of the egg and penetrate it head first as shown here.

Left Many sperm are necessary to break down the egg's outer coating and without such disrobing, fertilisation cannot take place.

egg's thick outer covering. Each sperm's store of protein-dissolving enzymes is then released, allowing its head to burrow downwards towards the egg cell itself, a journey taking about 30 minutes. Many sperm take part in this disrobing process, helping the eventual victor on its way. When the first sperm encounters the egg cell, the outer membranes of the sperm and egg fuse, and chemical changes in the egg rapidly prevent any more sperm from entering. If two sperm do achieve fertilisation, such as can happen if the egg is 'over-ripe', the embryo will not develop beyond an early stage.

Once the sperm is inside it releases a chemical signal that activates the egg. The sperm's head is then drawn towards the nucleus of the egg, and their genetic material fuses. This is the moment of fertilisation, when the genetic make-up of a new individual is created. Will it have its father's blue eyes, its mother's wide feet? Will it be prone to heart disease? The blueprint for all these hereditary characteristics is laid down now. The eventual sex of the child depends on whether the sex chromosome of the fertilising sperm was an X (for a girl) or a Y (for a boy). Because females have two Xs for their sex chromosome pair, all eggs carry X and males therefore have the pairing XY. The way in which embryos differentiate into male and female is explored further in chapter four, 'Puberty'.

The mother-to-be is entirely unaware of these happenings, but once they have occurred she is in the business of creation. Or rather, another human being is creating itself, exploiting the mother much like a parasite. Initially, on the fertilised egg's three or four day journey down the Fallopian tube to the uterus, it is quite independent. Every half day or so it divides. As each baby contains billions of cells this early replication might seem leisurely, but the sequence 2, 4, 8, 16, 32, 64 and on (and on) quickly gives rise to colossal numbers, reaching over a million after only 20 divisions. When it gets to the uterus the cell cluster starts absorbing fluid to create a central space inside it. Two to four days later it implants in the uterine lining, burrowing down using

Right Only one sperm will penetrate the egg. As soon as it does so there are rapid changes in the egg's membrane which are intended to prevent other sperm from entering.

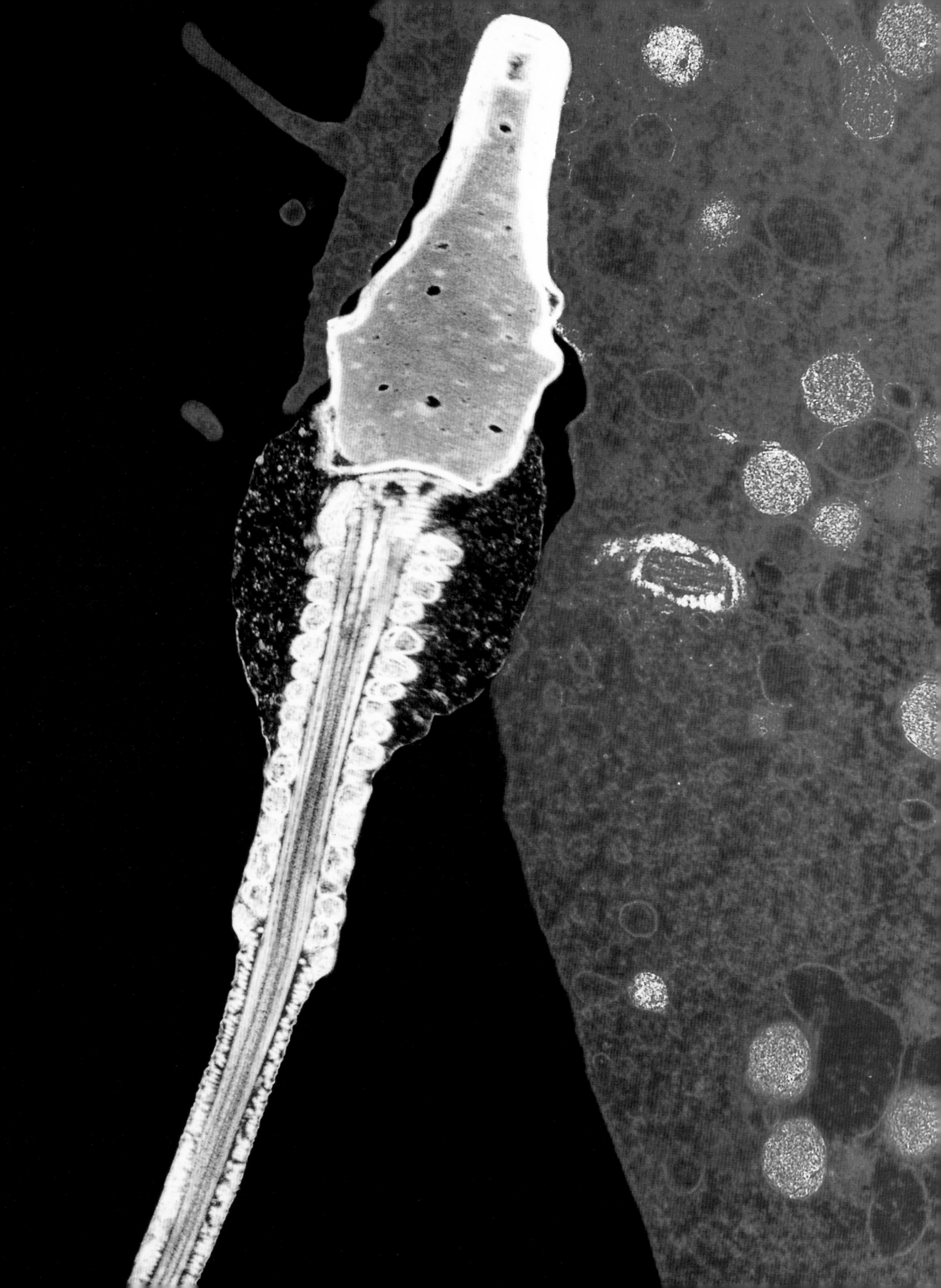

The testicle is a seething hive of constant activity. Not for this organ the leisurely production of one or two reproductive cells a month for perhaps 40 years of life. The testicle churns out its product by the million, from puberty until death. But despite the difference in scale, the genetic events that produce the egg and the sperm are similar, so a detailed look at the male system gives an insight into the female one also.

Each testicle contains about 900 seminiferous tubules, coiled structures half a metre in length. These are lined with a layer of continuously dividing cells called spermatogonia. As they divide, one of the daughter cells takes its place as a spermatogonium while the other begins a series of divisions that eventually leads to the formation of 16 sperms.

As human body cells, the spermatogonia have 46 chromosomes, made up of two paired sets of 23. One set comes from each of the man's parents. During sperm production the cells divide in such a way that each daughter cell is left with only one chromosome set, that is 23 chromosomes. Some of these 23 will be derived from the man's mother and some from his father, in any one of over four million possible combinations. Within each combination there is even more variability, because when chromosomes of the same pair touch each other they swap DNA through a process of breaking and recombining. Thus the fixed genetic combination of each individual can give rise to an essentially infinite number of combinations in his sperm.

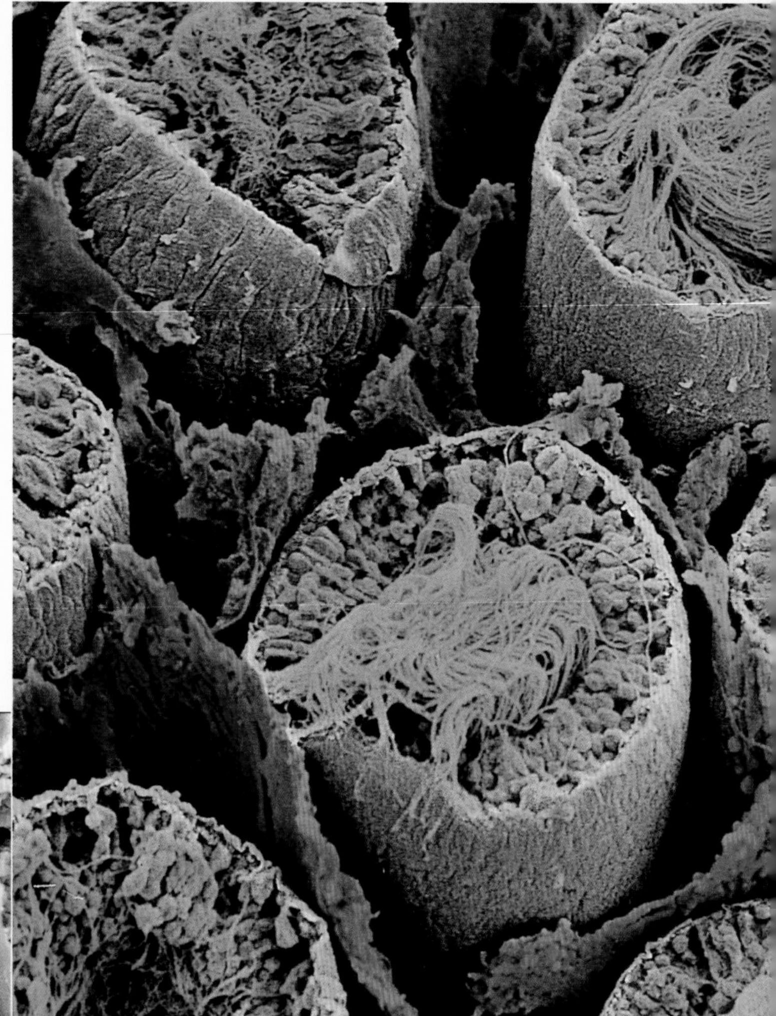

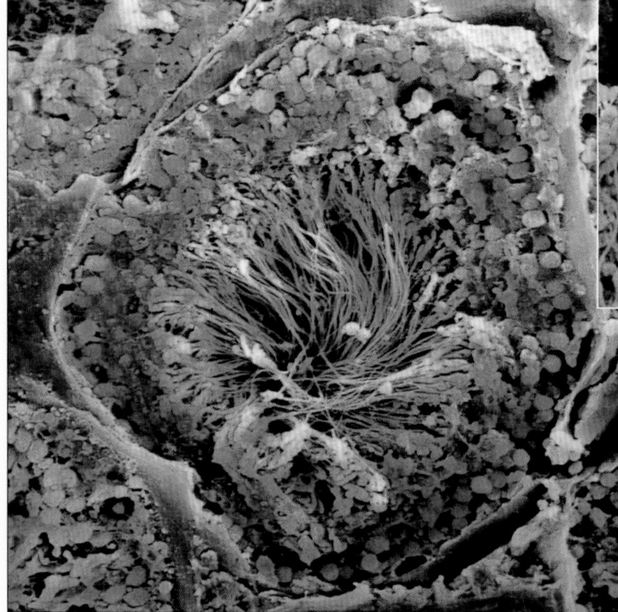

● *Above* Each sperm takes about 70 days to develop. The process starts at the outside of the tubule. The sperm in the centre have developed tails and will soon move on to the testicle's storage area.

● *Above right* Sperm are formed in the seminiferous tubules of the testis. The developing sperm cells are shown in blue. Surrounding the tubules are the cells that produce testosterone (seen here as dark orange).

A similar process occurs when eggs are made in the ovary, but here each egg-producing cell generates only one ovum, because each time the cells divide one of the daughter pair shrivels away. The crucial point is that, uniquely among human cells, the egg and sperm have only one set of chromosomes. Only when they combine at fertilisation is the genetic complement made complete again, creating a

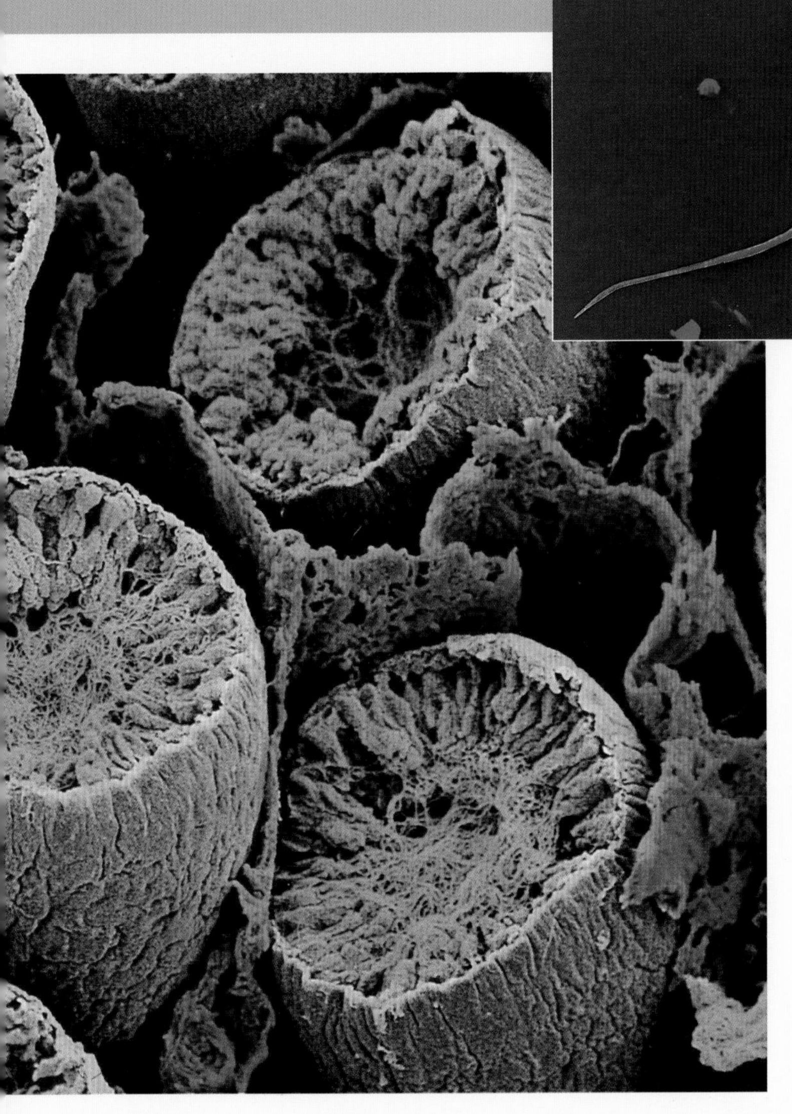

● *Above* Even young and healthy men produce some abnormal sperm. If the proportion of abnormal sperm is too high the male will be infertile, despite having a huge number of healthy sperm.

blue print for a new individual different from any the human race has ever seen before.

Once cell division is complete the sperm must undergo a lengthy process of nurturing and maturation before they are capable of action. As they work their way to the inside of the tubule they elongate to form a head and a tail. The head is given a cap carrying the enzymes the sperm will need to break into the egg, and the tail develops the energy supply to enable it to swim. Bathed in nutrients and hormones secreted by other testicular cells, the sperm are gradually collected in a 6-metre-long tube called the epididymis, where they are stored. A young adult man produces about 120 million sperm each day.

Whilst in the man's body the sperms' activity is suppressed by chemical inhibitors. They can survive in the epididymis for at least a month, but without a mechanism to propel them towards the egg they are useless.

At ejaculation the sperm pass from the epididymis through a tube called the vas deferens into the duct of the penis, where they are joined by the milky fluid from the prostate gland and various other secretions to make semen. Semen contains a cocktail of chemicals, nutrients and hormones to protect, activate and feed the sperm as they begin the race that only one will win.

Ejaculation is the result of an irresistible buildup of nerve impulses, starting in the genital area and passing up the spinal cord to an unknown area of the brain. Impulses also arrive from elsewhere in the brain due to visual stimulation or sexual thoughts and associations. First the signals trigger an erection, and then their intensity is increased during intercourse by stimulation of the nerve-rich tissues of the glans penis. This eventually triggers the release of semen into the penile ducts, a sensation that brings stimulation to its peak and causes the rhythmic contraction of the ducts and pelvic muscles that expel the sperm at orgasm. The ejaculate contains around 200–400 million sperm, but these account for less than 10% of its volume.

The huge numbers of sperm produced may seem wasteful, but they appear to be necessary. Men who produce fewer than about 70 million sperm per ejaculation – a number that would seem more than adequate – are likely to be infertile. Other infertile men have a normal sperm count but a high proportion of deformed or non-motile sperm: observation under a microscope reveals sperm with two heads, two tails or twisted shapes. Sometimes the sperm look normal but are unable to move properly. Strangely, men with a high count of abnormal sperm are usually infertile even if they appear to have plenty of normal sperm in their semen as well.

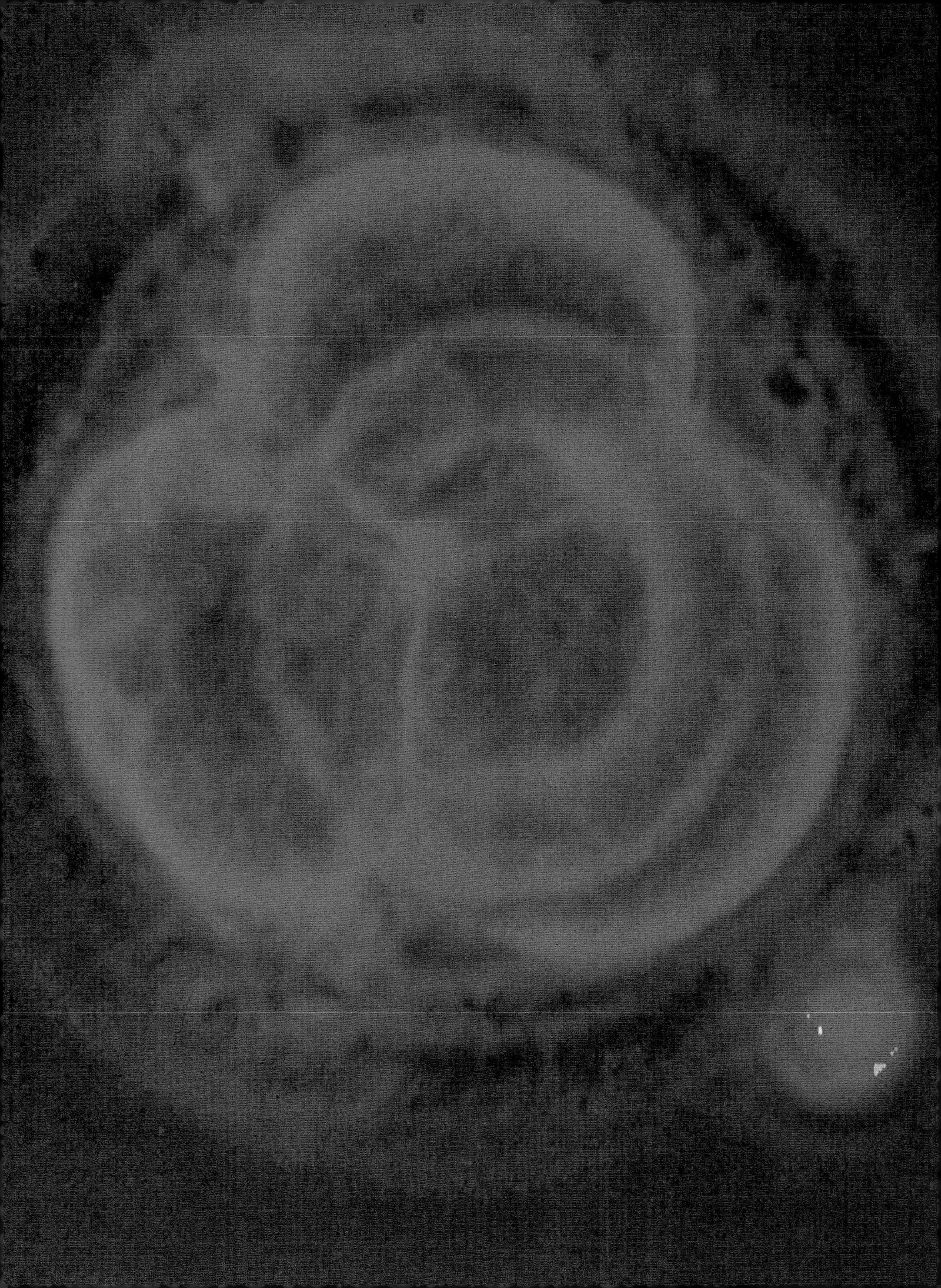

enzymes that digest the surrounding cells. The fluids and nutrients released are absorbed by the embryo and used as nourishment. If all goes well, the tiny invader will be securely implanted by six to seven days after conception. All the time it is secreting hormones, signalling to the mother's body that she is pregnant.

A large proportion of fertilised eggs fail to implant, and are lost with the menstrual flow without anyone realising that a conception took place. Occasionally the embryo implants outside the uterus, usually in the Fallopian tube. This occurs once in every 100–200 pregnancies and is known as an ectopic (out-of-place) pregnancy. As the embryo increases in size it causes severe pain and eventual rupture of the tube, leading to a potentially fatal haemorrhage. In very rare cases the egg misses the opening of the tube altogether and is fertilised in the abdominal cavity. Even here it will occasionally continue to develop, and in a handful of women an abdominal pregnancy has been carried to term and resulted in a live birth.

Left The dividing embryo. The first cell division may not occur until 30 hours after fertilisation, but further divisions are more rapid, every half-day or so. The 4-cell stage (*main picture*) is therefore reached after a couple of days, well before the mother knows she is pregnant. The inset shows a 16-cell embryo.

The early embryo

Even for embryos that successfully implant, early pregnancy is a precarious time. Up to one in five pregnancies fails before eight weeks, usually because something has gone wrong and the new life is no longer viable. Perhaps there was a problem with the chromosomes in the sperm or the egg, or perhaps a mistake was made somewhere in the process of cell division and differentiation. Whatever the reason, early miscarriages act as nature's safety valve. In the vast majority of cases there is no reason why the same parents cannot go on and have a healthy baby in the future.

In the very early stages the developing embryo absorbs nourishment directly from its surroundings, using enzymes to digest the womb lining. The lining is hormonally prepared for this during the second phase of the menstrual cycle, its cells swelling with stored nutrients. But this situation cannot last. There has to be a better system for nourishing the embyo, and soon. On day eight another space appears in the cell cluster – the amniotic cavity, in which the foetus will float in a fluid-filled sac. Fewer than one in a thousand of the cells created by the end of the second week will go to form the foetus itself. The rest become the placenta, umbilical cord, the amniotic and yolk sacs and the membranes known as the chorion. By day 16 the placenta begins to provide some nourishment, but the embryo continues to rely heavily on the uterine lining for at least eight weeks.

The placenta forms the interface between mother and foetus, and is a joint creation. Finger-like projections of embryonic tissue spread deep into the uterine lining and form tiny blood vessels. At the same time, the maternal tissues react to the invader by surrounding it with blood vessels of their own. A dense network builds up, providing a huge surface area. Maternal and foetal blood are separated only by extremely thin capillary walls, allowing nutrients, gases and wastes to diffuse across. The blood never actually mixes, reinforcing the fact that mother and child are two entirely separate entities. Some animals are able to regulate the flow of nutrients to the developing young, but humans

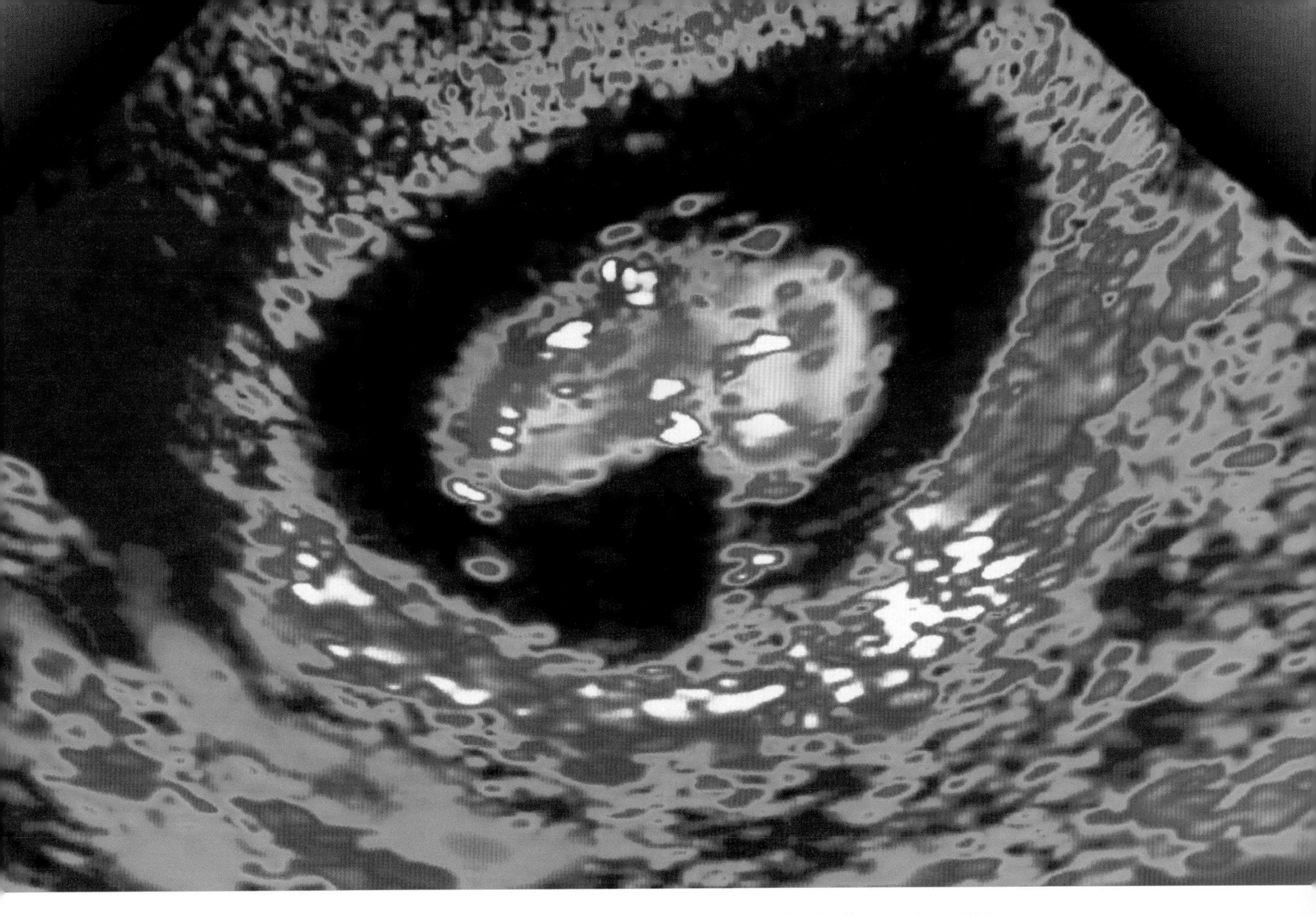

Above Ultrasound scan of a nine-week human foetus within its amniotic sac (black). Ultrasound uses high-frequency sound waves to reveal both internal and external structures, enabling doctors to check that the foetus is developing normally.

have no such control. If the mother's body is diminishing (perhaps from starvation) the foetus will still grow.

The life-giving placenta is discarded after birth without much reverence for its formidable achievements. It permitted nutrients to pass from mother to child – the essential supplies of glucose, amino acids, water and vitamins. It handed over oxygen and took away carbon dioxide. It served as a dialysis machine, removing waste products, and acted as a gland by producing hormones. Then, perhaps tugged gently by a midwife via its cord, it is abruptly rejected by the uterus and expelled as afterbirth. Ancient Egyptians, who respected this extraordinary object, used to carry it on parades. Many mammals, including herbivores, eat it, and even some humans now opt to consume it, loaded as it is with useful substances. Cosmetics companies collect placentas from hospitals for use in skin-care products. At the very least, the placenta deserves some admiration for the absolutely crucial part it has played in bringing the foetus to fruition.

A human being's development bears no similarity to the creation of, say, a car. Human equivalents to hose-pipes, plugs or wiring are not added as completed entities but start with a single cell. This cell multiplies and soon there is folding, folds being the first stage of many embryonic structures. Each such bulge or tube then enlarges and becomes specialised, and is soon quite distinct from the tissues around it.

Initially there is simplicity. There are three layers, much like a cake with filling in the middle. These three layers then curl to form a tube. The early embryo is like a worm, with a gut running from one end to the other, an outer covering also running from end to end and a central layer filling the space between the two. In humans (and all other mammals) that outer layer will form skin, nails, hair, eye lenses, mouth lining, tooth lining, salivary glands and all nervous tissue. The middle layer will make muscle, bone, cartilage, blood vessels and kidneys. The innermost layer will produce the oesophagus, stomach and intestine, as well as the liver, pancreas, bladder, and the lining of each lung.

There is a logic to this development. Skin, nails, hair and so-forth are all outside things. Intestines, liver and lung lining are on the extreme inside, and in the middle are muscle, bone and cartilage. The only oddity is that the nervous tissue, including the brain, comes from the outside layer. This occurs because, at a very early stage, a longitudinal fold forms along that layer's length and is pinched off to form the neural tube. This tube then becomes the spinal cord, and swells at the head end to form the brain.

The heart starts as a group of cells below the brain that somehow become electrically excitable. How this instruction is given exclusively to this specific cluster is a mystery, like so much of embryonic development. The group becomes a flat sheet of irregularly beating cells, which in turn folds over to become a tube. Why does it fold? Again, science has no answer yet. Only three weeks after fertilisation the twitching 'heart' can be detected. After four weeks the tube has developed a kink, and after five the separate left and right chambers have been formed. The foetal heart eventually beats about 150 times a minute, twice as fast as the mother's.

Twins

If twins are to develop instead of a singleton they can do so at several different times. Fraternal (non-identical) twins can only arise from the production and fertilisation of two eggs, an event more likely to occur in older women and when levels of one of the hormones involved in ovulation are high. Genetic predisposition to a high hormone level would explain the tendency for fraternal twinning to run in families on the maternal side. Its incidence also varies widely between populations, being extremely low in Japan and highest in West Africa, where it can sometimes reach 1 in 22 of all births. In the UK it is about 1 in 110.

Identical twinning is rarer and its incidence more constant, fluctuating worldwide between 1 in 260 and 1 in 340 births. Therefore in the UK about one in three twin births is of identical twins, while in Japan the proportion is nearer half. Identical twins occur when a developing embryo or fertilised egg splits in two, and they therefore share exactly the same genetic make-up. If the split occurs before the fourth day they will each have their own placenta and amniotic sac, as fraternal twins do. If it happens between days four and seven they will share a placenta but still have separate amniotic sacs in which to lie. If the split is later they will share both placenta and sac, making them less likely to survive than other identicals. Siamese twins, who share portions of their bodies, are believed to split after 13 days – but we cannot be sure if this is true.

Right The constriction apparent in this developing embryo may indicate that twins are on the way. If so, they will be identical twins as they are formed from a single embryo splitting into two.

What is certain is that many more twins are conceived than are actually born – the so-called 'vanishing twin' syndrome. This was first suspected as early as the 1940s but has been conclusively demonstrated using ultrasound scanning. Studies in which women are scanned regularly from the first weeks of pregnancy often show the initial presence of two embryos, one of which subsequently disappears. Sometimes two foetal heartbeats are detected in a mother who later gives birth to a singleton. The lost embryo is usually resorbed without symptoms, or else there is a little bleeding. However, the prognosis for the surviving twin is good. From these studies it has been estimated that one in six to one in eight pregnancies begin as twins, but only 1 in 50 of these twin conceptions will result in a twin birth. Put another way, 12–15% of all human beings alive start out with a twin, and for every set of twins born another 10–12 twin conceptions end as singletons. To put these figures into perspective, only about one fertilised egg in every four results in a live birth.

Multiple births usually arrive early. Instead of 38 weeks (for singletons) the average gestation times are 36 weeks for twins, 33–34 weeks for triplets and 30 weeks for quads. For humans, who already spend relatively little time in the womb, any reduction in the gestation period leaves them more vulnerable. The birth of more than two babies by natural means is very rare, although triplets sometimes occur when two eggs are fertilised and one then splits to form identical twins. In many ways it is remarkable that multiple births occur at all, given the disadvantages of sharing a womb. Some societies have even killed twins, resenting them as a form of abnormality and sometimes accusing the mother of infidelity. Blood group testing has proved that on rare occasions fraternal twins can in fact be sired by two different fathers, the so-called double mating when the sperm of two men are present in the woman's body at the same time.

Twinning can teach us much about human development. Identical twins reared apart help disentangle the effects of genetic inheritance and environmental nurture. Fraternal twins reared together have shared the same environment since conception but have different genes. Recently the presence of two

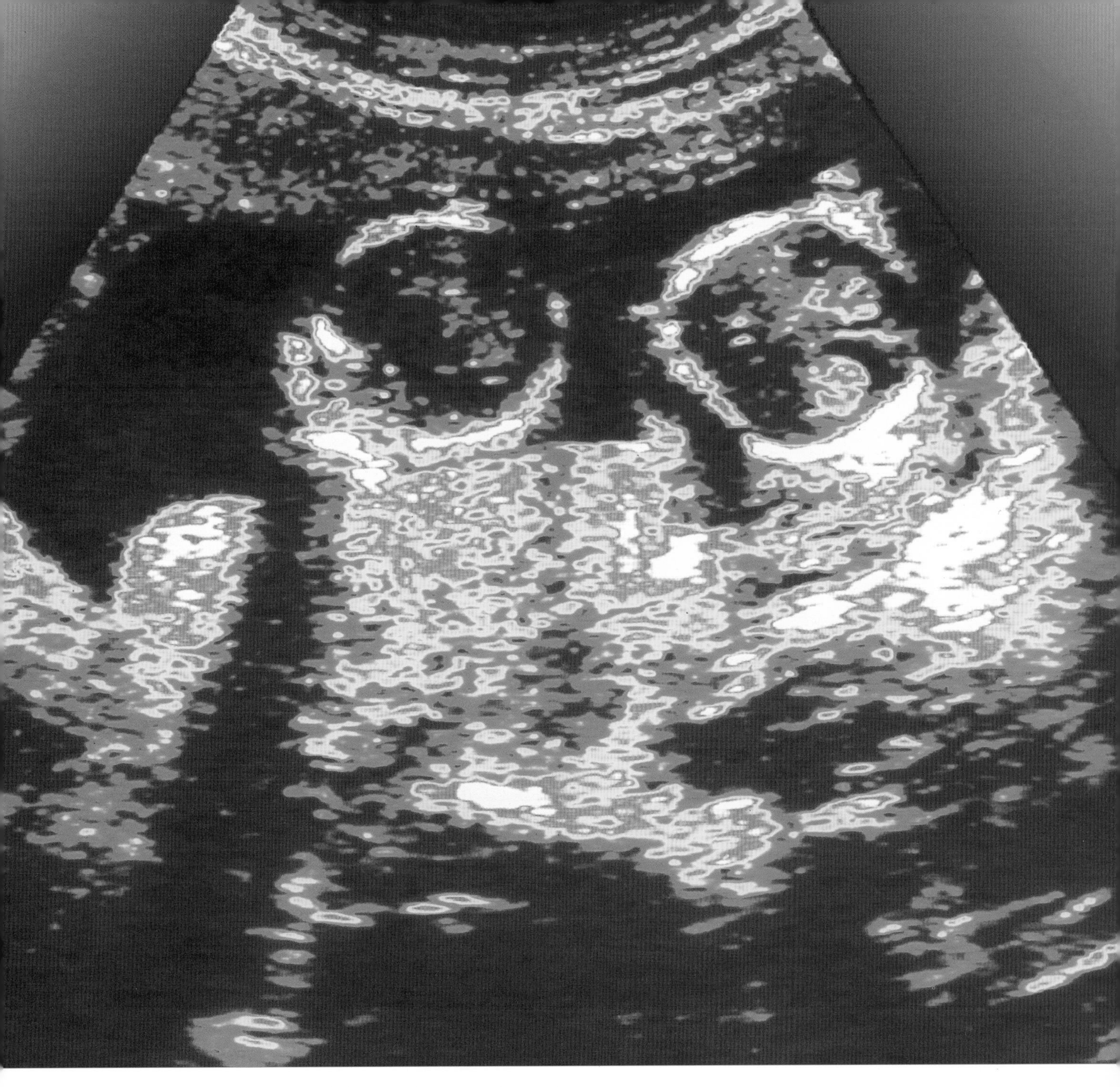

Above Twins in their mother's uterus, shown using ultrasound.

developing humans inside a single mother has led to a better understanding of foetal behaviour. Monochorionic twins (inhabiting one sac) have been seen to touch each other from 66 days (9.5 weeks), to react to a touch from 72 days, to kick each other from 73 days, 'box' with their arms from 80 days, 'embrace' from 86 days and 'kiss' from 92 days (13 weeks). Some of these actions come from individuals weighing less than a gram. Attempts at inter-human contact therefore occur long before these minute forms are capable of independent life.

THE FIRST EIGHT WEEKS

Development is a gradual process, but certain events become more conspicuous on certain days and can be said to begin, on average, at those times. The following calendar is a guide to the first eight weeks of a pregnancy. The first eight weeks are known as the embryonic period, a critical time of extremely rapid development when the beginnings of all the major organs and systems are formed. Timings given here are dated from fertilisation, so that birth would be due at week 38. The medical convention is to date a pregnancy from day one of the last period (LMP), so the due date is week 40 and the timings given here would have 14 days added to them.

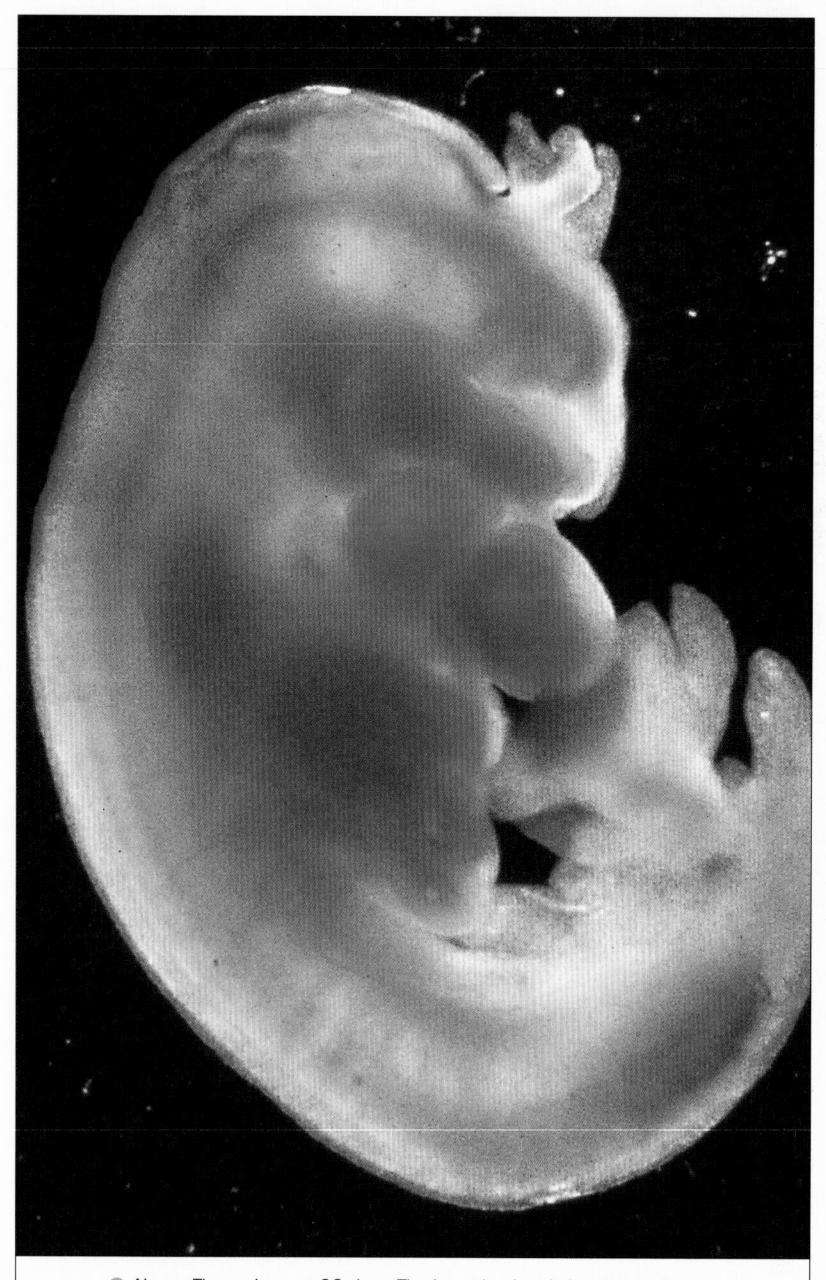

Above The embryo at 28 days. The heart is already beating and most of the organ systems are present in a primitive form.

30 HOURS

first cell division

DAYS

3–5 growing bundle of cells enters uterus

15 germ cells (for testis and ovary) developing

17 first sign of blood vessels

18 eyes and ears begin to form

21 heart tube constricts, the prelude to the formation of the heart chambers. The embryo is the size of a grain of rice

22 heart starts beating

23 head end bends sharply to form a bulge (no head structure before this)

24 upper limb buds appear

27 lung divides. At this stage the embryo is only 2.5 mm long, but 10,000 times larger than the original egg. Its umbilical cord is still rudimentary

30 heart pumping regularly
external ears begin development

31 arm buds differentiate into hand, arm and shoulders (the foot buds are still flat and less developed)

32 nose and upper jaw start to form

33 stomach, oesophagus and duodenum distinguishable

37 nose tip visible

38 eyelids start to form. Embryo is now about the size of a kidney bean

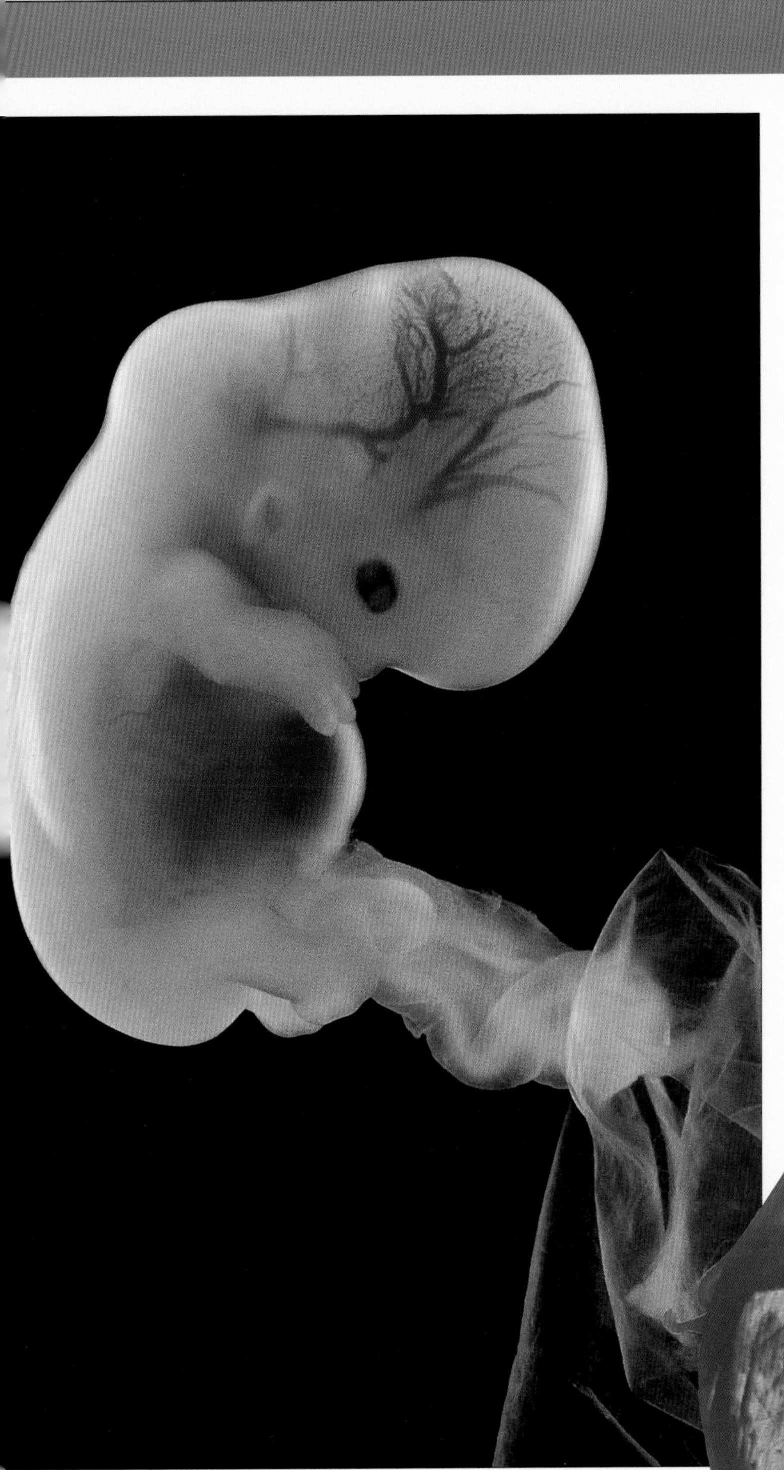

WEEKS

6 all major organs now present in rudimentary form
five fingers separated, but toes still only outlines
marked skeletal growth, but skeleton is still cartilage not bone
complete intestinal tube formed
male/female differences becoming visible

7 embryo now the size of a grape
stomach begins making digestive juices
muscular reflexes begin, but mother not aware of them
liver and kidneys start to function
outer and inner ear almost complete
jaws, lips and mouth clearly formed
teeth buds start
thumbs become different from fingers
first true bone cells

8 ankles, wrists, fingers and toes all now visible. From now on the embryo is known as a foetus (though the mother has probably called it her baby right from the start)

● *Above* At six weeks the embryo is recognisably human in appearance, and sex differences are now becoming visible.

● *Right* An eight-week embryo within its protective sac of fluid. It is linked to the developing placenta by its umbilical cord. From the end of the eighth week, the embryo becomes known as a foetus.

Life in the womb

Reactions to sound begin at around 16 weeks, before the ear is even complete. The uterine environment is never quiet, with the mother's heart thumping and the equally pervasive borborygmi – all the rumblings, gurglings, burps and murmurings of the mother's alimentary canal. However, foetuses seem to hear best at high frequencies, so these low-frequency body sounds are probably muffled. There are sounds from outside too. No one is sure of their exact impact on the baby, but some mothers have reported violent kicking in response to loud rock music. The baby certainly hears its mother's voice: at 28 weeks its heart rate will increase when she speaks, and newborns are able to recognise their mother's speech.

Life in the womb tends to be portrayed as a time of serenity, the foetus floating calmly in its protective sac. Reality is somewhat different. Not only is the uterine world far from silent, but the idea of gentle repose in the curled-up 'foetal' position has been shown through ultrasound scanning to be a myth. Only in the last eight weeks or so, when it has become too big to move around freely, does the baby spend all of its time curled up. By the ninth week, when only the size of a grape, the foetus has begun moving its muscles. By 12 weeks it has become a little acrobat, rolling from back to front, somersaulting through the amniotic fluid and waving its limbs as if to explore its surroundings. Even the muscles of the face are exercised with frowns and lip movements. The mother knows nothing of this, the foetus being far too small for her to feel. Not until about week 18 will she begin to feel fluttering movements inside. This event was known as the quickening, and until the modern era was considered to be the beginning of life. Pregnancy was not acknowledged until the baby could be felt moving. Luckily for the mother, the baby soon develops a pattern of waking and sleeping, confining its most vigorous movements to intermittent periods of strenuous activity.

Although it probably does not experience five clearly defined senses in the way fully developed humans do, the foetus does seem to have some awareness of taste and smell. Odours from the mother's food and drink permeate into the amniotic fluid and are thought to be absorbed through the nose and skin, allowing the baby to 'smell' in much the same way as a fish is able to. Children of chilli-lovers may experience their first spicy meal well before birth! From about 12 weeks the foetus begins swallowing amniotic fluid and urinating it out again. Observation has shown that it swallows most when the composition of the fluid is sweet, suggesting an awareness of taste.

Foetal perception of pain is a contentious issue. Foetuses react to amniocentesis needles from 20 weeks, but this could be reflex withdrawal rather than positive dislike of the stimulus. Nerves are not connected to the brain before 13 weeks, therefore pain before this time is highly unlikely. Neural connections begin at 13–16 weeks and are well established by 20, so amniocentesis needles could be more than irritating. High blood levels of stress hormones do occur when such foetuses are tested for response to possibly painful stimuli, and as stress hormones are produced by the pituitary gland in the brain there must have been messages relayed from affected skin to receptive brain. But does that amount to the awareness of pain by the foetus? One (not very satisfactory) answer is that foetuses do have, long before birth, a stress response to the kind of stimuli that would be painful to us. It is therefore prudent to assume that foetuses beyond 26 weeks at the latest *may* suffer pain, a possibility not even considered until quite recently. It was also assumed in the past that newborns were incapable of sensing pain, but this idea has been discredited. Those who sever newborn's foreskins without anaesthetic should be aware that the pain to the baby from such a sensitive area may be horrendous.

The stages of foetal development from nine weeks until birth are listed chronologically on pages 46 and 47. Generally speaking, the initial three months (first trimester) are dedicated to development and the subsequent six (the second and third trimesters) to growth. The first 14 days are mainly dedicated to preparation of the site. By 12 weeks, although it weighs less than a gram, the foetus is recognisably a human being, with all organs in place and some already working. It responds to tickling. Its heart pumps 28 litres (50 pints) of blood a day, and its kidneys produce urine. It swallows and digests amniotic fluid.

FROM NINE WEEKS TO BIRTH

By the end of the ninth week since fertilisation (week 11 if dating from LMP), the critical development phase is over and the risk of congenital abnormalities developing is much reduced from now on. The remainder of the pregnancy is dedicated to rapid growth in size and the further growth and differentiation of the organs and tissues developed during the embryonic period. The rate of growth of the head relative to the rest of the body slows down, and the baby gradually begins to assume the proportions it will have at birth. Growth is especially rapid at 9 to 16 weeks. Again, the following timings represent averages.

WEEK

9 sex now easily detectable
fingernails begin to form
the eyelids close over the eyes
non-identical twins begin to look dissimilar

10 mother has gained about half a kilogram (1 lb) in weight (though this varies greatly)
baby about 6.5 cm long and weighs 18 grams (half an ounce)

11 baby can move its thumb to its finger
the halves of the palate fuse and the vocal cords are in place

12 urine is produced and excreted into the amniotic fluid
amniotic fluid is swallowed and digested
all nourishment now comes from placenta
the form of the genitals is established
arms now almost at their final relative length
stroking of the lips causes a sucking response

13 skeleton begins to turn to bone
uterus moves upwards and out of pelvis

14 circulation is now effective, the heart pumping 28 litres (50 pints) a day

15 baby thought to be aware of loud noises from outside
head hair, eyelashes and eyebrows visible

16 'the quickening' – the mother begins to feel the baby's movement
thumb sucking begins
egg follicles formed within the ovary of the female
eyes and ears now close to final position; face looks more human
nipples visible

● *Left* At eleven weeks the foetus's organs are all fully developed. The rest of the time in the uterus is principally devoted to growth.

● *Right* Four-month foetus shown here with its umbilical cord. The outer ear and nose and mouth are formed but the eyelids are not yet complete.

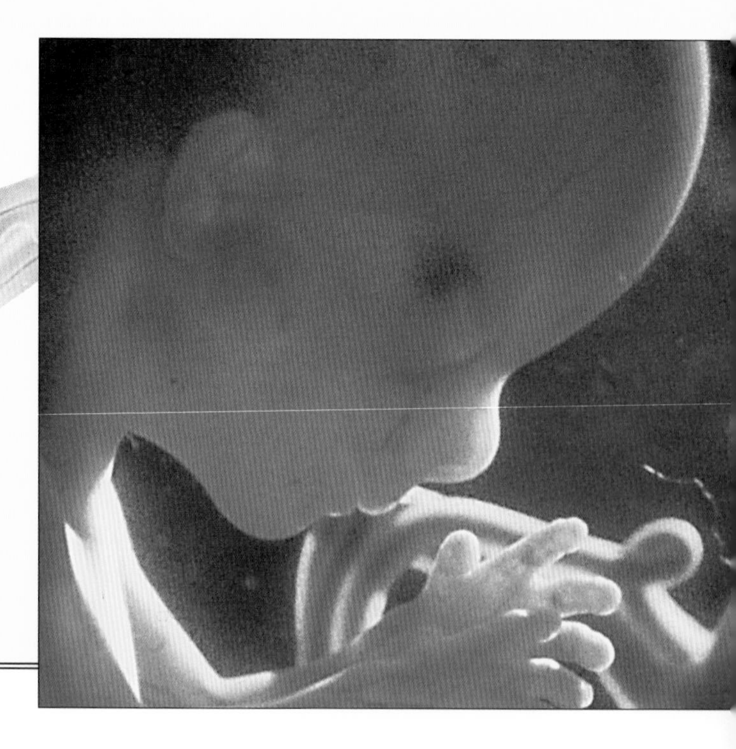

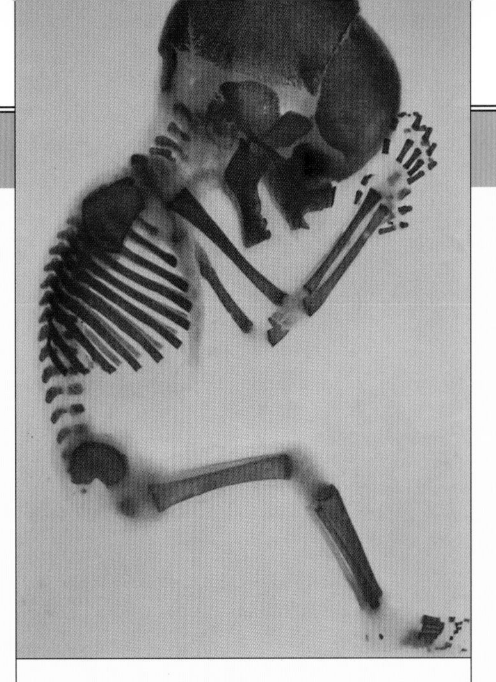

Above Bone development of a fourteen-week foetus. The foetal skeleton originally consists of cartilage which is replaced with bone up to and after birth.

WEEK	
18	baby is the size of an orange
19	the half-way stage: baby is half the length it will be at birth but only 10% of the weight. It has no chance of survival if born at this time. Mother has gained about 4 kg (9 lb) and is now gaining roughly half a kilogram a week
	the eyes open
	lanugo hair appears – fine, downy growth all over body
20	baby develops a pattern of activity and sleep
	arms and legs reach the final relative proportions seen at birth
	baby now covered with vernix – a mixture of sebaceous secretions and dead skin cells that, it is thought, may prevent skin chapping from exposure to fluid
	face becomes less wrinkled as fat is deposited
21	uterus may begin 'practising' contraction (known as Braxton Hicks contractions)
22	there is now a small possibility that the baby might live if born prematurely, but most die because their lungs are too immature
	hand will grasp strongly
24	half of babies born now will survive given sophisticated care, but many of these develop problems
	skin still very wrinkly due to lack of fat
25	80% of prematures survive given the right care

WEEK	
26	lungs now sufficiently developed – virtually all prematures now survive, though birth at this stage is undesirable
	thumb-sucking commonly seen
	fat beginning to be deposited
	lanugo hair begins to disappear
28	baby responds to its mother's voice
29	the three-quarter stage. White fat now accounts for 3.5% of body weight, and baby looks less wrinkled. It tends to stop its earlier gymnastics and settles into a head-down position. Mother has put on an average of 8.6 kg (19 lb); the placenta weighs half a kilogram
30–34	pupils show reflex response to light; baby is aware of a glow when the sun shines on the mother's abdomen
	skin is pink and smooth and the baby is chubby
35	baby orientates spontaneously towards light
	it begins practising breathing movements, and can get hiccups
	head (or presenting part) can drop into pelvis from now onwards
37	growth slows or stops
	white fat now accounts for 16% of bodyweight
	baby is about 50 cm long; males grow faster and weigh more at birth than females
	mother has gained 12.2 kg (27 lb); placenta weighs nearly three-quarters of a kilogram
38	birth. Only 5% make their 'due date', and 8–12% wait another 2–3 weeks to emerge

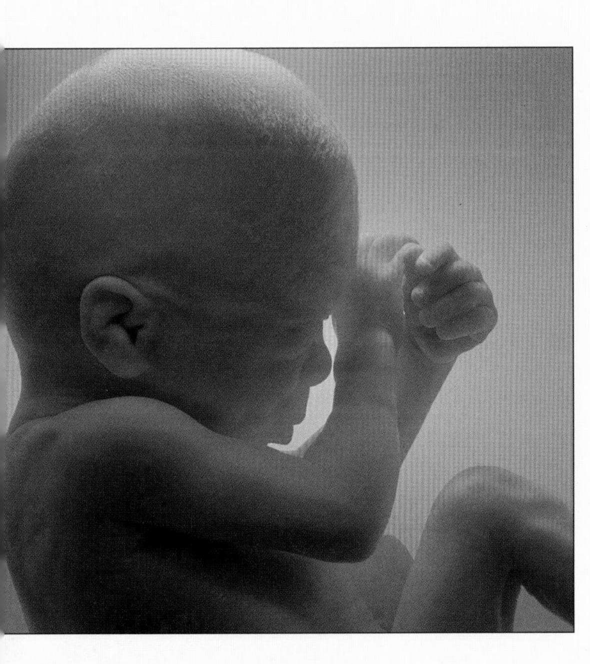

Left A foetus at five months. The features have not filled out, as fat will be laid down in the final weeks.

The first three months, while the organs and nervous system are being formed, are the time of greatest susceptibility to harmful agents such as drugs, pollutants, certain infections and excessive alcohol.

The pregnant woman

Right from the first days after fertilisation, the dividing egg sends out hormonal signals to tell the mother's body she is pregnant. The ovary continues secreting progesterone, so that instead of being broken down the lining of the womb carries on thickening. Other hormonal changes kick in, so that shortly after their period is missed many women are already beginning to feel pregnant. The blood vessels in the breasts begin to swell, making them tender and uncomfortable. The nipples may become darker and more prominent, and the breasts get larger. Over 70% of women experience nausea (which may or may not lead to vomiting) in the first three months. Many describe it as like feeling sea-sick. It is often worst in the mornings when the blood sugar is low, but can strike at any time of day. Sometimes it is triggered by particular foods or smells: pregnant women often develop strong aversions to foods they normally eat without a second thought.

Pregnancy sickness seems to be universal; the idea that it only occurs in Western societies is a myth. It was documented by the Egyptians as early as 2000 BC. It has been speculated that sickness is a way of protecting the foetus from harmful toxins during its most vulnerable period. It is most prevalent in the first 12 weeks, precisely the time when all the major organs are being formed and when toxins would do the most damage. The placenta, with its protective filtering action, is not yet fully functional. It is thought that the mother's brain lowers its threshold for monitoring harmful substances in the bloodstream, triggering nausea and vomiting more easily.

Right This woman's pregnancy is close to full term. To cope with the demands of pregnancy and birth, the mother's body undergoes many physical changes brought about by alterations in the levels of certain hormones.

Whatever its purpose, sickness is often considered a sign of a well-established pregnancy. Women who experience it may have less risk of miscarriage than those who do not. In the past, and in parts of the world today where there are no pregnancy tests, it has always been one of the first indications to a woman that she is pregnant.

Opposite to food aversions are the cravings some pregnant women experience. These may be for anything, from raw carrots or peanut butter sandwiches to earth and lumps of coal! Perhaps cravings are a way of ensuring that the foetus obtains the vitamins and minerals it needs. Some cultures believe that cravings and aversions are the baby speaking through its mother, asking for the foods it wants.

By the end of the first three months, the sickness and tiredness of early pregnancy have usually passed. The mother's body has become used to the invader, and for a time the two co-exist happily. From about weeks 11 to 30 the amount of blood in the mother's circulation increases steadily, to cater for the demands of supplying nutrients and oxygen to the baby. Her heart must work harder and harder to pump the blood around her enlarged body and through the placenta – its output is up 20% by the 15th week. Her chest expands (separately from the breasts) to accommodate deeper breathing and the enlarged heart. Hormones begin softening the body tissues, allowing them to stretch and preparing them for the birth.

The breasts too are being prepared. Until a woman has been through a pregnancy her breasts are not capable of lactating, as the required structures are not present. From the earliest weeks after conception, hormonal signals cause the system of glands and ducts to extend and develop. Stores of fat and protein are laid down, from which the milk will be made. By the fifth or sixth month, the breasts are ready to commence their feeding role.

As the baby grows the demands on the mother's body increase, and for the last three months her physiology is stretched to the limit. The expanding uterus presses against the internal organs, squashing them together. The mother's heart is actually twisted around so that it lies on its side. Acid heartburn is common as the stomach becomes constricted, and pressure on the bladder

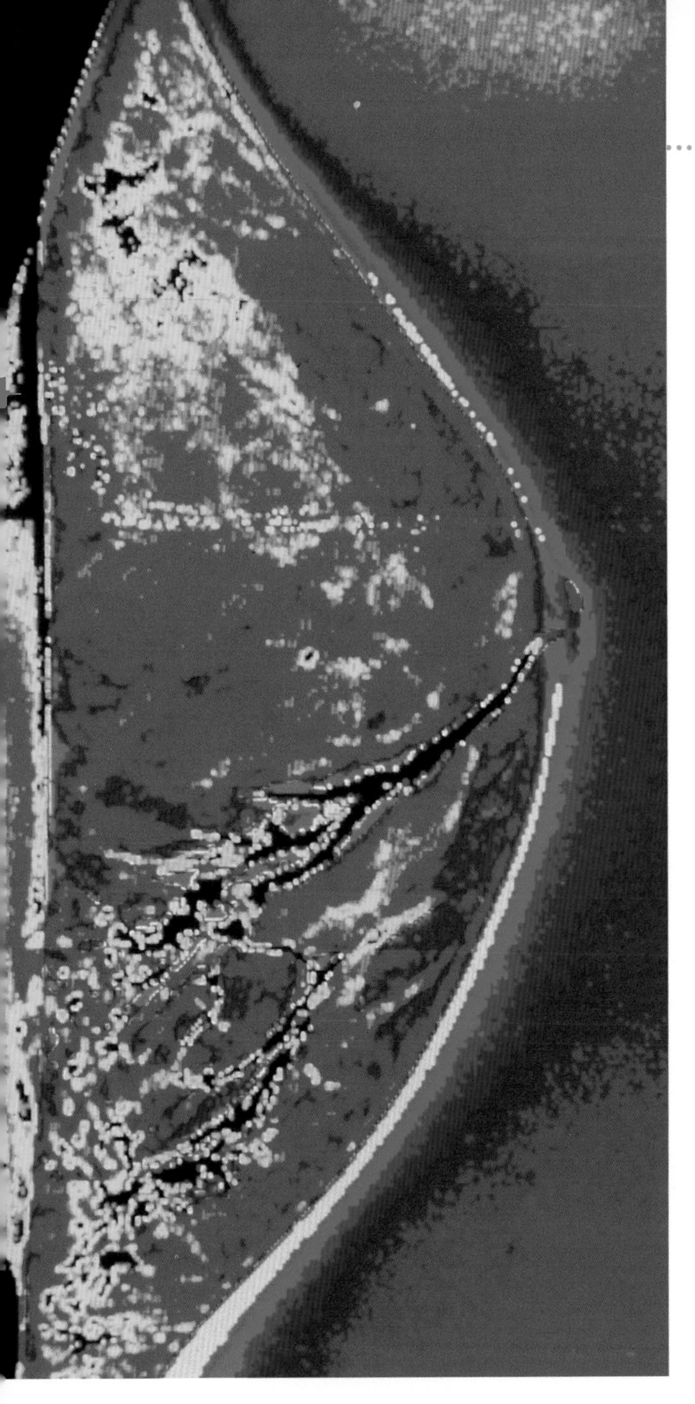

Above Hormonal signals cause milk glands and ducts to develop in the breasts during pregnancy. In this X-ray image the milk ducts appear in black.

brings a frequent need to urinate. Backache is a common complaint as spine and muscles work to support the growing bulge. Stretch marks may appear as the skin is pulled out of shape. Tiredness reasserts itself, and by the last month or so many women are thoroughly fed up with the pregnancy and long for it to be over.

The large increase in the mother's weight after conception is not fully understood. On average she will have gained 12.2 kg (27 lb) since conception. The placenta will weigh 0.7 kg, the amniotic fluid 0.8 kg, the uterus 0.9 kg, and the baby itself 3.3 kg. Her breasts will have increased in weight by 0.4 kg and she will have 1.4 kg more blood. These gains total less than 8 kg, and nobody is quite sure where the rest comes from. Fat gain and fluid retention are the most likely explanations.

Some time in the six weeks before birth the baby's head becomes lodged between the pelvic bones, where it will stay until delivery. Its phenomenal nine-month rise from cell cluster to well-developed baby completed, it is poised to begin the shortest but most momentous and dangerous journey it will ever make.

CHILDHOOD

Every year 280 million children are born. All over the world parents happily anticipate the years ahead when their child will travel a unique route along the well-worn path from infancy to adulthood. In each case, the brand new offspring is emphatically a dependant, almost a form of parasite. Yet each helpless being has the determined capacity to amend, drastically, the lives of others. On its own behalf it will cause great quantities of time, energy and cash to be spent catering for its needs. Its growth and progress during the preceding nine months have been phenomenal, but the months and years ahead will be no less dramatic or exciting, and will be even more demanding as babyhood merges into childhood.

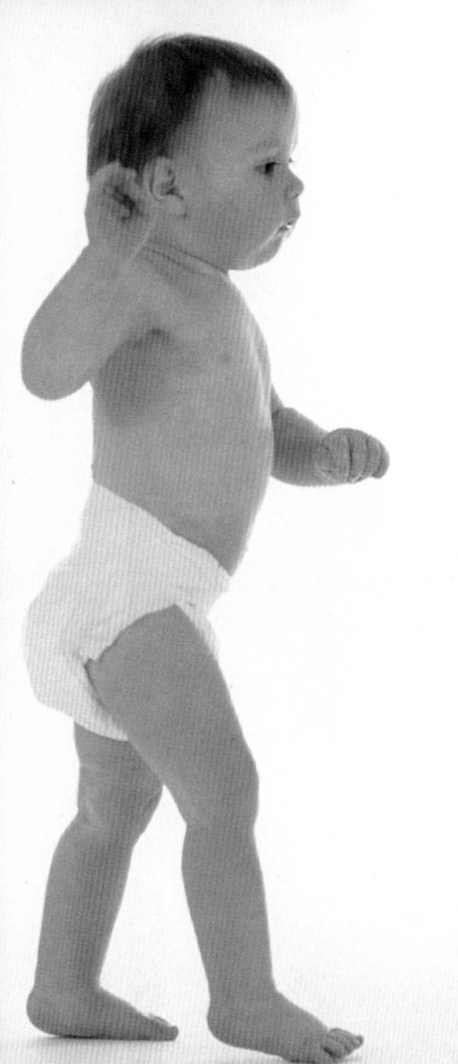

A risky time

Humanity, it has been said, is very good at lots of things, but having babies is not one of them. In nineteenth century Europe, urban infant mortality was about 200 per 1000. One-fifth of all babies therefore never reached their first birthday. Greater prosperity and improved public health brought gradual improvements – the figures for Britain are typical of developed nations.

Left Part of the organ of balance (the vestibular apparatus) in the inner ear (*main picture*). The red crystal at the bottom forms part of the 'ear dust'; these crystals help humans adjust to gravity. Without a sophisticated system of balance such as this, a child could not learn to stand and walk bipedally (*inset*). Although the impulse to walk is instinctive, the acquisition is through learning.

JANE AND RICHARD

For Jane and Richard, an extremely important baby has just arrived in the world. Holding him for the first time Jane is overcome with the wonder of it all. She hands her new baby to her husband. 'He's lovely. Here he is. Have you got him?'

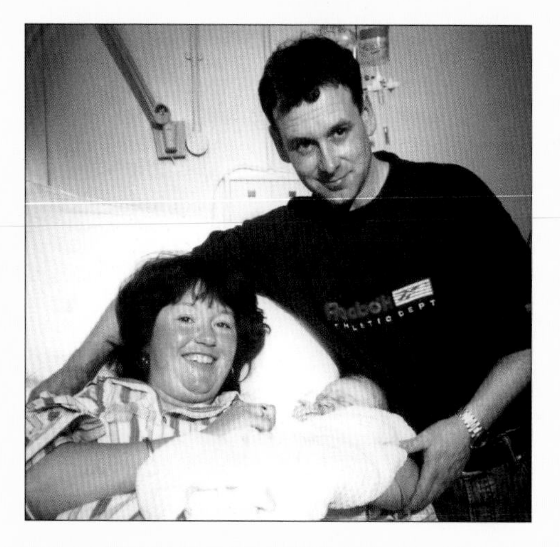

Richard takes his son. 'He feels nice and warm. Like a hot water bottle. Worth all the effort?'

'Definitely,' says Jane, 'definitely. Aren't we lucky? His skin is so soft, so soft. He's so kissable.'

We've all done it – just the once. We've all decided the time has come to change from the cosy aquatic world of our mother's uterus to the bigger world outside. First we have to get out and our biggest obstacle to this is the size of our head, which with a circumference of 35 centimetres is even wider than our chest.

Inside our mother's body we have already manoeuvred ourselves into a head-down position; at least, most of us have. But there's difficulty right from the start. Our head is longest from front to back, so we have to get ourselves sideways for the best fit. As our mother's contractions shove us down the birth canal, we have to rotate through 90 degrees to fit into the available space. At the same time, to make our heads a fraction smaller, our skull bones start to overlap under the pressure. At some point the fluid sac breaks and there is a gush of water. The stress is enormous, overwhelming. There's more heaving and shoving, and then – at last – it's all over and we slip out.

We gasp. Our lungs expand. They are abruptly filled with air. And we yell, probably. And someone cuts the cord that has been our life-line. And the noise is tremendous. We have overcome the most momentous journey of our lives. We have been born, just one of the nine human beings born every second of every day.

In 1904 the district of Salford, near Manchester, recorded mortality of 128/1000 for breast-fed babies, 263/1000 for those fed on cow's milk and 439/1000 for those given condensed or sugared milk. By 1928 the figure for Britain as a whole had improved to 60/1000, though in 1935 the Great Depression caused an upward blip to 62/1000. In 1955, after seven years of universal free healthcare under the National Health Service, it was down to 27/1000. In 1992, sophisticated medical care had brought mortality down to just 6.4/1000 for boys and 5.8/1000 for girls.

There is always likely to be a hard core of deaths at birth time, mainly due to congenital abnormalities. Most of the infant mortality reduction this century has resulted from better care of children aged between one week and 52 weeks. The first seven days are now the most fraught in developed nations, with more British children dying in that first week than in any subsequent year until those same individuals reach their thirties and forties. The developing world still witnesses the old style of child mortality. More than 11 million children under the age of five die worldwide each year (20 a minute), with pneumonia, diarrhoea, measles, malaria and malnutrition killing 70% of them. These five illnesses are all largely preventable. The worldwide infant mortality rate of 121/1000 is similar to the British rate during the earliest years of the current century.

Human births are difficult, because of the large foetal head and the narrow pelvic girdle (relative to apes). The pelvic alteration was caused by becoming upright. 'Lucy', the 3-million-year-old hominid fossil from Ethiopia, already had a narrower, less rounded pelvis than her ape ancestors. But even for other mammals, with their relatively smaller heads and more accommodating pelvic bones, birth is not a simple affair. The switch from liquid to gaseous environment, from placental nourishment to lungs and stomach, is always problematic. The time is still dangerous, and death is still possible. Nevertheless, human births are exceptional, and the passage we all take (except for Caesareans) through the 10 centimetres of birth canal has been called, with reason, the most dangerous journey any of us will ever undergo.

An early problem during birth is related to the baby's head shape, which is longest from its nose to the back of its skull. Because of this shape the baby is forced to enter the constriction of the birth canal by facing sideways. The canal then changes its shape so that its longest dimension is from front to back. The baby has to follow suit, and rotates through 90 degrees. Further on the canal changes yet again, being broader at the front. The baby's head is wider at the back, which means that in a normal birth (95% of the total), the baby actually appears with its face pointing downwards and away from its mother. (Pot-holers may understand this best, as they too twist and turn to manoeuvre their shoulders – an adult's widest part – through restrictive openings.) The circumference of a baby's head continues to be bigger than its shoulders for its first six months.

The pressure exerted on the baby, the total expulsive thrust, is about 11 kilograms, and has even broken obstetricians' finger joints. The baby is undoubtedly stressed at this time. Levels of adrenalin are extremely high in the newborn, higher than in adults during a heart attack, but they are beneficial. Adrenalin

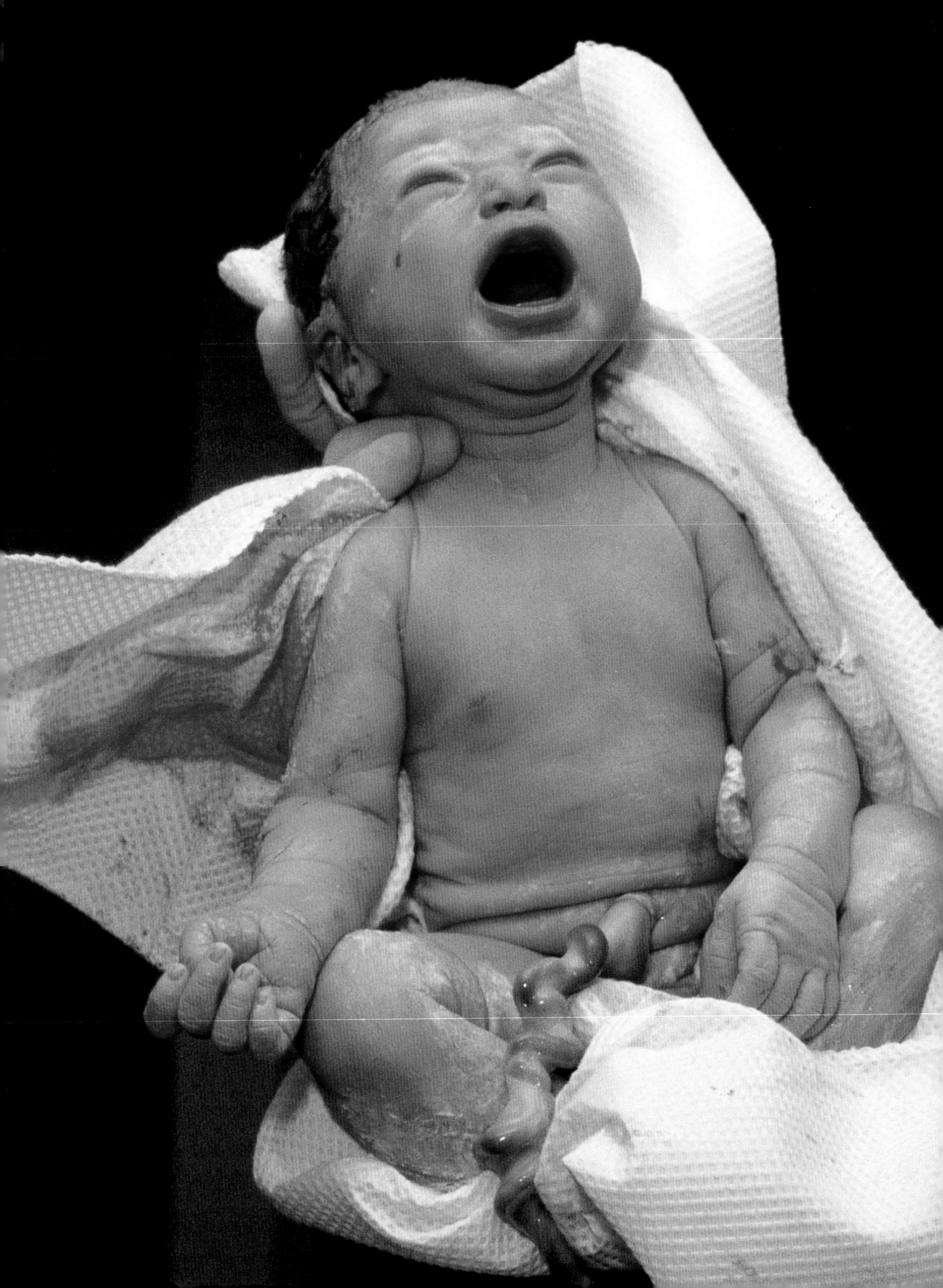

helps to force the baby to take its first gasps, to expel most of the amniotic fluid from its lungs and reabsorb the rest, to inflate the lungs with air, to divert blood their way. In short, to start a brand new style of existence.

Before birth the baby's lungs are collapsed, filled like the nose and mouth with amniotic fluid. Oxygen and waste products are supplied and removed directly to the bloodstream via the umbilical cord, bypassing the lungs altogether using a special foetal configuration of veins and arteries. When the source of oxygen is abruptly switched to the lungs, a major replumbing of the circulatory system is necessary. The foramen ovale, a valved opening connecting the two chambers of the heart, is slammed shut by pressure changes within the heart. The closure becomes permanent in the next few weeks as the flaps of tissue fuse to form a solid wall. The new pattern also causes the closure of a duct between the pulmonary artery to the lungs and the aorta (the artery to the rest of the body), and of another duct in the liver. This is the equivalent of major heart surgery, yet it all goes on in the first few minutes of life.

The first breath requires a colossal effort to fight the surface tension that keeps the fluid-filled lungs closed. It is like pulling apart two surfaces of a collapsed balloon filled with moisture. To reduce the surface tension and prevent the lungs collapsing when the first breath is exhaled, the lungs contain surfactant chemicals. These are produced quite late in pregnancy, causing problems for premature babies born before the surfactant is in place. Lung immaturity is often the critical factor that limits the survival of the very premature.

Most of the amniotic fluid is expelled through the nose as breathing is established. The first gasp for air will use about half the potential lung space. The alveoli, the grape-like gas exchange surfaces, will be fully deployed by about the third day. Breathing at 30–80 times a minute (compared with an adult's 15 times) and taking in 20 cubic centimetres of air per breath (instead of an adult's 500 cc), the baby's respiration is then working effectively.

Left A one-minute old baby girl – breathing air. The baby must rapidly adapt to acquiring its oxygen from the air using its lungs instead of being fed all it needs via the umbilical cord.

A baby's pulse rate is much higher than an adult's, being on average 123–126 beats per minute on day one, rising to 130 at two months, and then dropping to about 120 by the first birthday. An adult's heart beats about 70 times a minute when at rest. In contrast, blood pressure is much lower than in adults, rising from about 67/37 on day one to 76/44 on day four, and 99/57 at the first birthday. A young adult's blood pressure is nearer 120/80.

A pressing problem facing each newborn is temperature control. Foetuses within the uterus have no such concern, being maintained at maternal heat. Almost inevitably the outside temperature they encounter is cooler than the 37°C of the uterine incubator, and newborn babies tend to cool in consequence. Drops of 1.5°C have been recorded in the first hour, despite room temperatures of 27–29°C. Babies are rarely able to shiver, an important means of generating heat for adults. They have less insulating fat than adults and possess a large surface area in relation to their bulk, relatively larger than at any later stage in life. But around the neck, shoulders, breastbone and spine lie deposits of brown fat, a substance found in many young mammals. When the baby is cold its body begins to burn the fat via a unique metabolic process that produces high levels of heat. A good bout of crying, or any form of exercise that increases metabolism, is also warming (provided the baby is adequately clothed).

Premature infants are even more vulnerable, as their fat reserves did not have time to develop *in utero*. Babies kept in heated incubators tend to be left unclothed, partly for better observation of their body colouring but also because garments actually prevent ambient heat from reaching the enclosed individual. A skull cap is sometimes used, the scalp being large in surface area and therefore an important avenue of heat loss.

Newborn babies are also poor at keeping cool, with their sweating system imperfectly developed for the first few months. In any case babies have slightly higher temperatures than adults, with 'normal' rectal temperature at three months being 37.4°C, at three years 37.2°C and at 13 years 36.5°C. Despite poor heat regulation babies soon acquire the pattern of daily alteration, with temperatures (as for adults) highest in the evening and lowest at dawn, the diurnal difference between high and low being a degree or more.

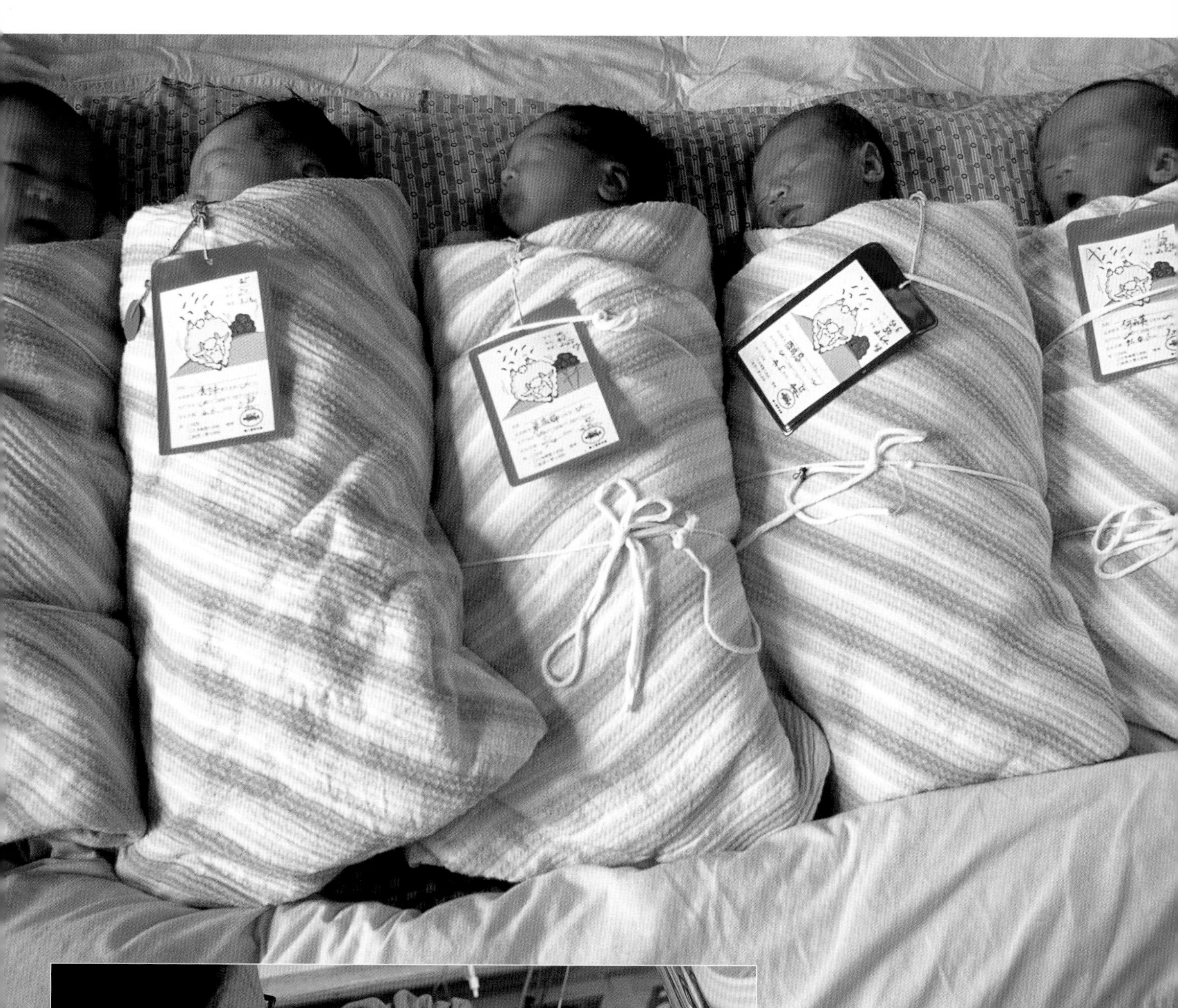

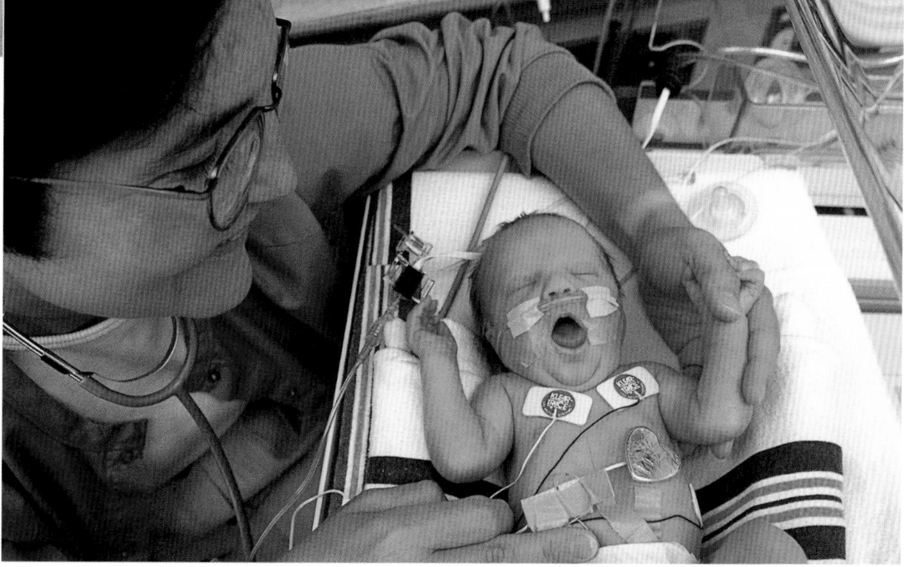

Above Swaddling was commonplace in Britain until the end of the 18th century, and is still common in areas such as the Balkans, Russia and Japan. It is said to have a calming effect on the babies.

Left Newborn babies, particularly the premature, are poor at regulating their temperature. Heated incubators are used for premature babies as a half-way stage between womb and cot.

Becoming independent of a maternal womb also means increased exposure to infection. Babies acquire antibodies against specific diseases from their mothers, but the amniotic sac was a haven from infectious assault. The world outside, however cleanly kept, is not. The mother continues to supply antibodies through the breast milk, helping ward off disease until the baby's own immune system takes over.

Bacteria within the gut can be both harmful and beneficial. For example, vitamin K has two forms, vitamin K1 and vitamin K2, the second produced by intestinal bacteria within a few days of birth. If the mother has provided insufficient vitamin K for her infant during pregnancy, the gut bacteria will soon make good the deficiency. Bacteria also synthesise several members of the vitamin B group. If the bacteria are killed, for instance by antibiotics, then vitamin B deficiency diseases such as beri-beri may occur.

A baby's skull bones are soft at birth. The process of birth, with all those contractions forcing the head to prise open a passage, causes a moulding and distortion of the head bones (structures which, at this stage, are not bones at all but connective tissue and cartilage). The bony plates – perhaps a better term – may be forced to overlap, creating small external ridges on the baby's head (and also amazement that, with so crucial an organ as the brain lying below them, there is no internal damage). There are six fontanels or 'soft spots', the one on the top of the head towards the front being the most conspicuous. It pulses visibly and even enlarges for the first two months of life, but thereafter shrinks, as do all the others, eventually closing between four and 26 months.

Odd-shaped heads will acquire a more normal form, and the ridges will disappear, within a week or two. Nevertheless the plates stay soft, and moulding can continue after birth, even caused by the head resting on bed-clothes. Such flattened skulls can last for a few months before symmetry is finally acquired, and the expanding brain may even assist in this correction. A baby's brain more than doubles in weight in the first year – 350 grams to 910 grams. By the age of three about three-quarters of the post-natal brain growth has been achieved. By age four it is thought that a child has the same level of brain activity as an adult. As Leo Tolstoy wrote: 'From the child of five to myself is but a step,

● *Above* As illustrated by this family group, the skeleton's proportions change through childhood and puberty.

● *Below* An X-ray image of a 2-year-old boy's hand (*left*). The five metacarpals (within the palm) are still short, whereas the 14 phalanges (of the fingers) are almost completely formed. Only 2 bones of the wrist have ossified (hardened into bone), and six more will develop later. By the time the child is six (*below right*), the bones of the hand have reached more-or-less their adult form.

but from the newborn baby to the child of five is an appalling distance.' It is indeed, and in many respects children are young adults by the time they start school.

In terms of size, the brain reaches more or less its adult dimensions at about ten years. Monkey brains, in comparison, achieve a greater proportion of their growth before birth. Even the chimpanzee, the most intelligent of the primates, completes its brain growth by the age of one year.

The proportions of the skeleton itself will also change in the post-birth years. The head, about a quarter of the body's length at birth, gradually becomes smaller in its relative proportion. At first there is hardly any neck, and there are high shoulders above a round chest. From 3–10 years the neck lengthens, the chest broadens, and the ribs slope increasingly downwards. Children's arms become 'normal' in appearance long before legs do: legs can look distinctly odd for several years. This mirrors foetal development, when the front limbs evolve most quickly.

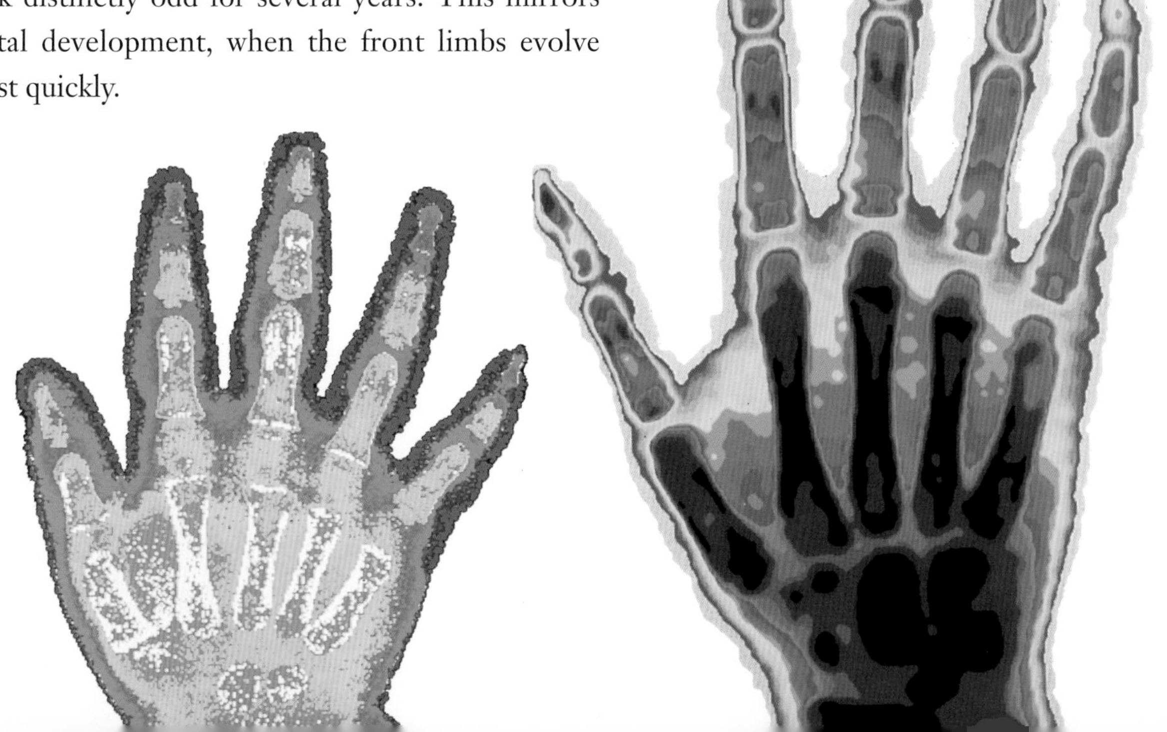

To be born or not to be?

By many yardsticks human babies are born too early. They cannot run on day one, as can antelopes, or even struggle gawkily to their feet like a newborn foal. Young monkeys can cling tightly to their mothers from the moment of birth (and may even clasp the maternal fur while being born). Many reptiles have to fend entirely for themselves from the moment they hatch, their parents being absent. It is intriguing to wonder at what age a human infant might survive if similarly abandoned (within a suitable neighbourhood of available food): four years, six, or more?

All birth is an evolutionary compromise. What is most effective for that species? To be born sufficiently competent to move with the herd at once, probably meaning a lengthy pregnancy, or to be born within some protective den at an early and relatively incompetent stage? The human brain is already sizeable at birth, with this organ and the rest of the nervous system forming one-tenth of the total body weight (as against one-fiftieth for an adult). A baby's brain will increase in volume 2.6 times during its first 12 months. Therefore its total weight will almost treble before the first birthday. Body length, 50 centimetres at birth, will become half adult height at 24 months. A child is then quite proficient, not exactly running with the herd and far short of an adult's mental prowess, but well ahead of its ability at birth.

To be born with all the developmental benefits of a two-year-old could be advantageous, but there is no question of such an enormous infant, with such a head, with such a length and weighing 9 kilograms, being carried and then delivered by a human mother – unless both she and her birth canal were a completely different shape and size. Weight and body length would be problematical, but head circumference even more so. Its measurement, even for nine-month gestation babies, is 35 centimetres. For babies older by one and two years the circumferences are 46 cm and 48.5 cm respectively. Thereafter head growth slows tremendously. From age two to seven the increase in circumference is approximately 0.5 cm a year, and from seven to ten just 0.3 cm a year.

The existing compromise of 3 kilogram babies with 35 cm heads has much to be said for it, despite the newborn's imperfections. Its increase in head circumference during the first post-birth year is almost double the increase from age one to ten. It would therefore seem (however much mothers may dispute the fact) that head size is kept deliberately small before birth in order to make birth easier, or rather to make it possible at all. In any case the human baby, even if born later, would still require many years of assistance and attention. The burden of parental care, for what Desmond Morris has called the naked ape, is heavier than for any other species. A few months less care would make little difference, but later birth would entail considerable amendment to the female body.

Life as a newborn

So what can a human baby do on day one of its external life? It can suck (at any object resembling a nipple), and can continue to breathe even when sucking/drinking (an ability lost by six months when the larynx assumes the adult position necessary for speech). It can swallow, salivate, cry (but without tears), smell, taste, hear, respond to touch, yawn, hiccup, sneeze, stretch, cough, and of course breathe, digest, excrete, and carry out all the other essential internal happenings. It also has some surprising extra capabilities, which are described on pages 64 and 65 in the box on reflexes.

It is a shame (or perhaps a blessing) that we cannot remember the earliest months, or even the initial years, of our lives. We might then better understand a baby's thoughts and needs. It can certainly hear, and will react to loud sounds as if hand grenades are exploding. By three or four weeks the baby will respond favourably to speech, finding it interesting (if incomprehensible). It can see from day one but fuzzily, partly because each eye's lens is not yet properly controlled by its musculature. Vision is clearest within a range of about a metre, hiding the world of unknowns beyond but ensuring that the faces of mother

EARLY REFLEXES

The newborn baby is a bundle of reflexes, some of which are short-lived and may reflect our evolutionary past. Perhaps the most striking is the grasp reflex, by which the baby grips tightly any object placed in the palm of its hand. For the first few weeks the grip is strong enough to support the baby's entire weight; its relative strength gradually decreases as the baby gets heavier. This ability is thought to be a vestige of the young ape's need to cling to its mother's fur. It even lingers on in the foot – stroking of the sole causing the toes to curl as if trying to grip. After two months this is lost and the toes then turn upwards and spread out when given the same stimulus. At two years the response changes again and the toes turn downwards.

Another ancient response is the rooting reflex, where the head, mouth and tongue all turn towards any contact with the baby's cheek. This helps locate the nipple. Later a protrusion of the lips accompanies the rooting. From the first, the baby will reflexly suck at any nipple-like object placed in its mouth. Other reflexes include sneezing in response to nose irritation or even bright light; and walking, where a baby held vertically with its feet on the ground will attempt to walk if moved forward. This disappears after six weeks, and is thought to be an early manifestation of brain pathways that will come into play later on.

There are various other oddities such as the crossed extension response: stretching one leg and stroking its sole should cause the other leg to bend and stretch. This goes after one month. These early reflexes must be lost to allow the conscious brain to take control and develop voluntary movement. Grasping any object placed in the hand is a good strategy for the newborn, but would be highly dangerous and disadvantageous in later life. Failure of the reflexes to disappear is usually a sign of developmental problems. However, many of them remain buried somewhere in the primitive part of the brain, reappearing occasionally in people with senile dementia.

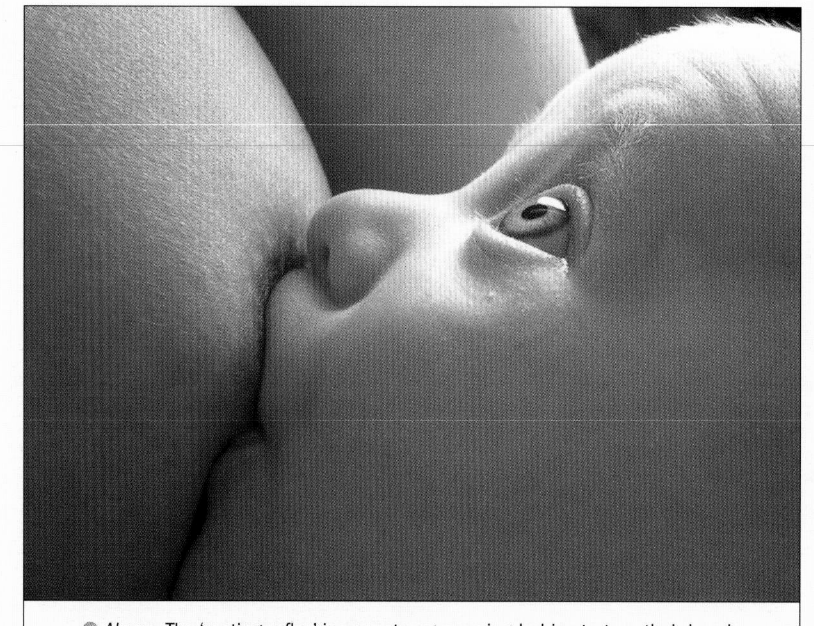

● *Above* The 'rooting reflex' is very strong, causing babies to turn their heads in the direction of any pressure on their cheek. However, newborn humans often seem clumsier than baby animals at actually locating the nipple.

● *Right* The diving reflex. This baby appears happy and relaxed underwater.

Another reflex seems so unlikely that for years no one was aware of its existence. A baby suddenly immersed in water will seal off its lungs and begin to paddle and kick. The mouth may be open, but the epiglottis (cartilage and tissue which covers the larynx during swallowing, preventing food 'going down the wrong way') diverts water away from the lungs. The urge to breathe is suppressed, and immersed babies appear content in their watery environment. This 'diving reflex' disappears after six months or so, just when babies start to breathe through their mouth. Once it is gone, immersion in water can cause extreme distress. But for our first six months, even before we can crawl, most of us are more comfortable underwater than we will ever be again.

and carers are clear. This probably prevents the newborn from becoming too anxious. Focusing steadily improves and, at two months, the eyes will follow a moving shape, such as a face. At four months there is sufficient coordination between eye and limb to reach for objects.

The simple act of voluntarily reaching out and grasping an interesting toy represents a triumph of coordination. Once learned it becomes automatic, but the baby is like a novice asked to fly a jet aircraft. Everything must be done by trial and error. Early attempts at reaching objects involve waving the arms and legs around until by chance contact is made. These uncoordinated searching movements are known as frobbing. Gradually the brain begins to remember the movements that led up to a successful action, and before long they can be reproduced at will, and the baby has mastered another skill.

Right The baby's mouth is not merely a place for food intake but for testing and experimentation. It forms part of the sensory experience – what does an object feel like? What does it taste like? Is it a form of food?

Master manipulators

The first smile is a treasured moment for parents, but initially the smile is a reflex and not a sign of pleasure at seeing a familiar face. That reaction comes later, often partnered by the chuckling vocalisations which are such a joy to hear. Neither is the first smile an imitation of expressions made by adults: even blind children will smile at four to six weeks of age.

However helpless they seem, babies are skilful manipulators of those around them, particularly their parents. Even the newborn exhibits manipulative behaviour, albeit unconsciously. Human adults have an inbuilt attraction to baby-like creatures that transcends culture. Research testing responses to different photographs has characterised the key features as a large head in relation to the body, a large forehead, large eyes set low in the face, rounded protruding cheeks and a plump, rounded body. These attributes are considered 'cute' and attract a protective response. The attraction is not confined to babies: consider the love for wide-eyed kittens and puppies, not to mention numerous cartoon characters devised to these very specifications. Perhaps the early reflex smile is another device to ensure the parents bond with the child and thus continue to supply its needs.

Even the mother's physiology can be manipulated by her offspring, whose cries trigger the release of milk by the breasts. For the first year, crying is the most important signalling system the baby possesses. To an outsider its cries may seem to be just a constant racket, albeit one that is difficult to ignore (manipulation again). Both parents and childless adults react to cries with raised blood pressure and heart rate, classic signs of anxiety. But the mother quickly comes to recognise different types of crying that convey different messages. Analysis of sound spectrographs has quantified what she has learned, revealing distinct patterns for the birth cry, cries of pain and cries of hunger.

Eating and sleeping

A baby's appetite is considerable in relation to its body weight. In the earliest days it may take in 3% of its weight per day, but this quickly increases to in excess of 10%. By ten days, with much slurping and sucking, it will, on average, have consumed its own body-weight in milk. By then it is putting on some 20 grams of weight a day, or about 1 gram for every 20 that it consumes. An adult, if consuming similar percentages of its body-weight, would be processing seven or more kilograms per day.

Newborn babies do not, as frequently alleged, spend all their time either feeding or sleeping. Sleep may last in total for an average of 16 hours a day, but can be much more or much less. Some normal infants will sleep for twice as long as other normal infants. One study discovered that average sleep time for 2.5-year-olds was 12.9 hours, but the lengths ranged from eight hours to 17. Babies have already established a sleep rhythm of their own within the uterus. As infants sleep for less and less time as they get older, it is assumed that near-term foetuses sleep for an even longer proportion of each 24 hours than do newborns. A newborn baby cannot help going to sleep (much to the annoyance of mothers who have bottle or breast all ready). Within a year the child has some measure of control over the timing.

Sleep has been divided into four stages, varying from light to deep. The lighter stages, also defined as REM sleep (standing for rapid eye movement), are experienced more by infants than by adults. REM sleep accounts for at least half of sleeping time for the newborn but 20% or less for adults. As there is a steady drop in such light sleep from birth to adulthood, pre-birth babies are thought to experience even more of it than newborns. REM sleep is linked with dreaming, a fact only verifiable for individuals able to report their dreams. It is intriguing to wonder if the not-yet-born and the very young do actually dream. If so, it would be fascinating to know the subject-matter of such early visions, given the dreamers' lack of experience of the world around them.

The developing brain

Before birth, the foetal brain develops from a primitive fold of nervous tissue into an organ that orchestrates the functioning of every system in the body. Amazing though this transformation is, we now know that it is only half the story. Shortly after birth the brain begins a second, explosive phase of development shaped not by pre-determined genetic programming but by the inputs it receives from its new world. The dense forest of interconnecting nerve cells, or neurones, can be compared to the vegetation of a jungle, with new growth thrusting forward towards the light and plants whose light source is cut off withering back. Just as plants respond to light, the neurones respond to sensory messages – sights, sounds, smells, a new reaching movement – and alter the way they connect with each other. The bombardment of new information reaching the baby's senses stimulates its brain into a constant process of rewiring, so that all the time it is becoming better able to make sense of the world and to control its own body.

Initially the brain seems to create chaos. Neurones in the 'higher' centres – areas concerned with language, sensory perception, voluntary movement, reasoning and emotion – begin a furious process of establishing connections to other cells. By the age of two each neurone has about 15,000 synapses or connection points, creating an incredibly rich network of potential brain pathways. Some of these will be reinforced by sensory inputs, and those left unused will wither away. Thus the structure of the brain is physically changed by the nature of the signals it receives. If the information coming in is inadequate – through lack of stimulation or faults in the sense organs – development can be seriously and permanently affected.

For example, the visual pathways in the brain and retina are not immutable at birth but are moulded during the first few years of life. Two-thirds of babies who are long-sighted at six months will have visual problems such as a squint or a lazy eye at the age of four. By the time these are picked up at school it is often too late to correct them, the faults having become

TEETH

An obvious milestone in early life is the arrival of the first tooth. Julius Caesar, Richard III, Louis XIV, Napoleon Bonaparte were born with a visible tooth, as are one in 2000 of more ordinary individuals. The normal date for its arrival is after 7.5 months (for which breast-feeding mothers must be relieved), with two-thirds of children cutting their first tooth between 5.5 and 10 months.

Teeth are formed from the skin, the enamel originating from the outer layer or epidermis and the inside of the tooth from the dermis. The milk teeth begin to form in the foetus at around seven weeks post-conception. Ten weeks later they are fully developed in the jaw, and the buds of the permanent

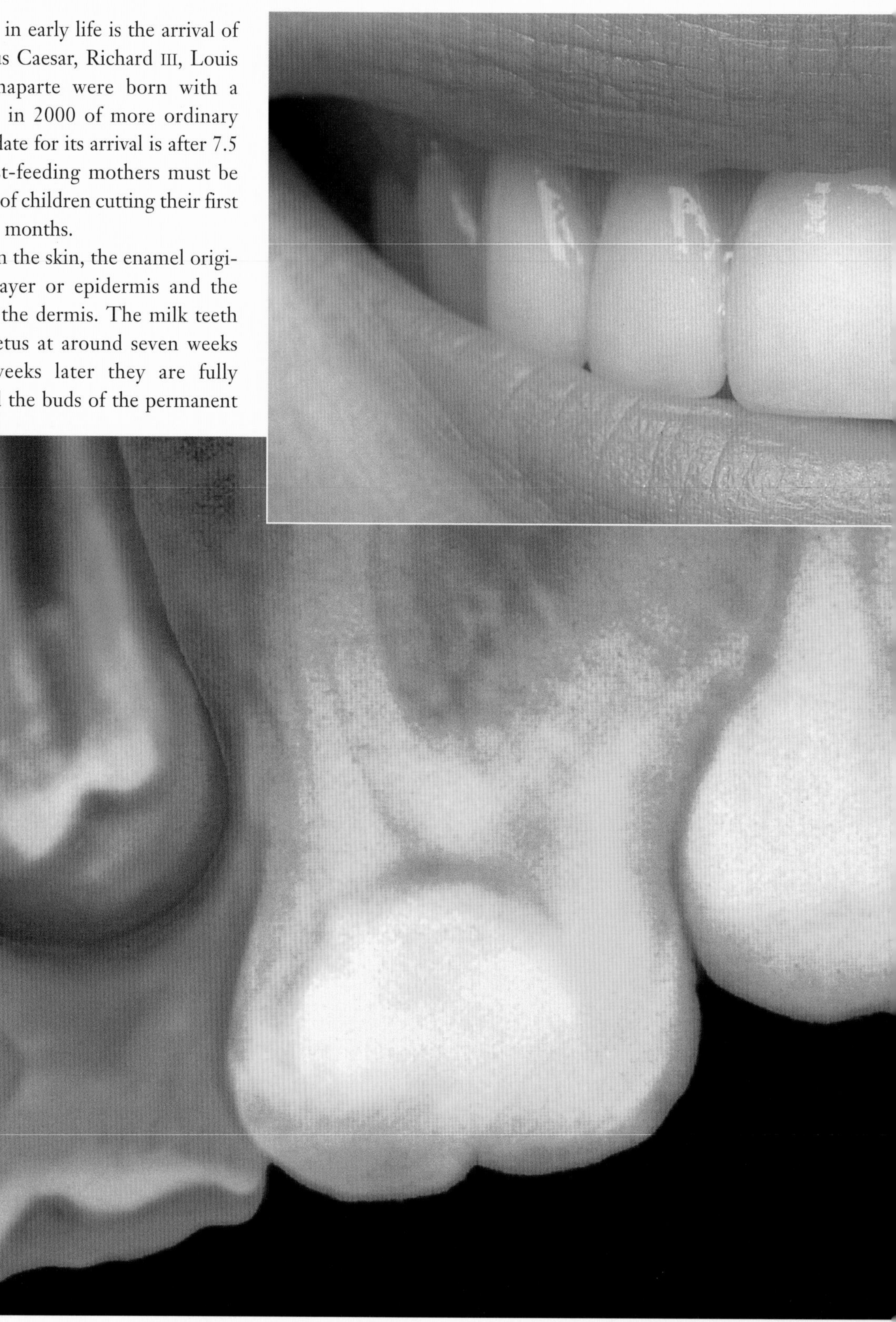

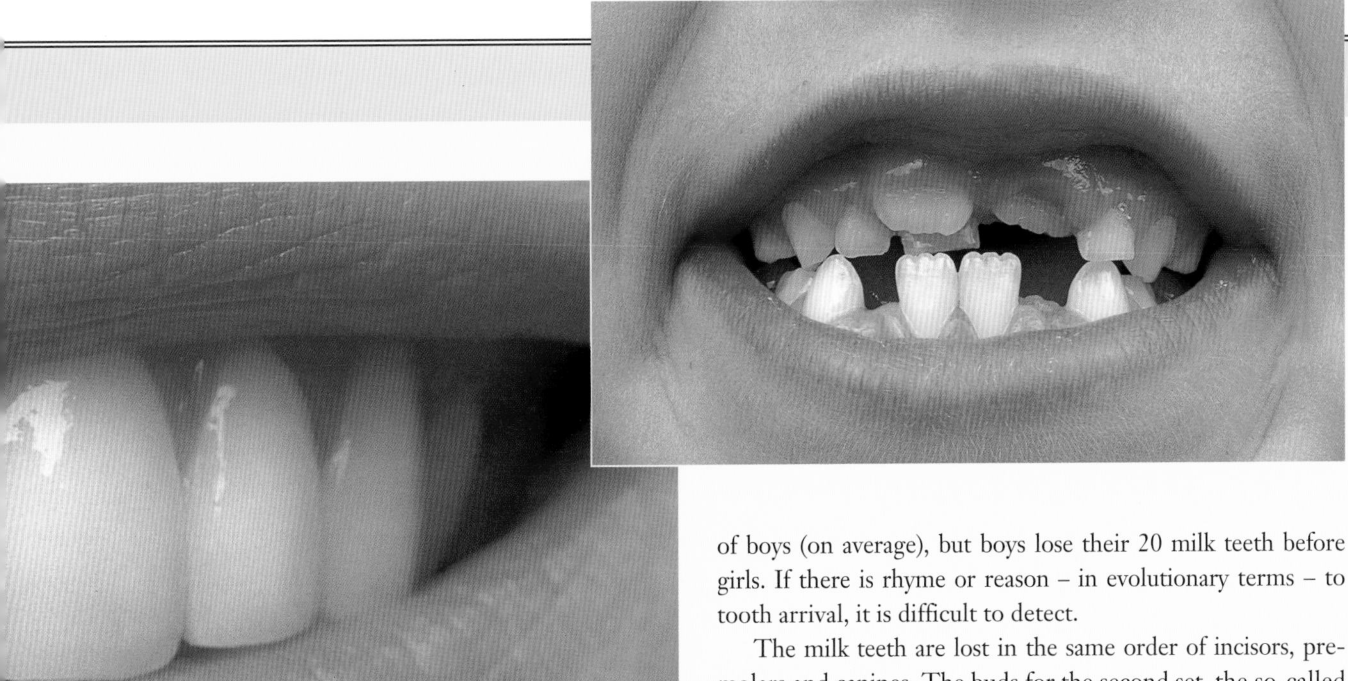

● *Below left* The coloured X-ray shows a permanent tooth (green) erupting under a child's milk tooth. The roots of the milk tooth are being reabsorbed, and it will soon become loose and fall out.

● *Above right* A few weeks later the permanent tooth will emerge, a process that can be seen in this seven-year-old's mouth.

● *Above* All being well, the apparent disorder will eventually be replaced by an orderly array of 32 permanent teeth.

teeth have started to form behind them. Each tooth consists of a protective outer layer of enamel, a middle layer made of dentine (a substance slightly harder than bone), and an inner core of dental pulp which carries the nerves and blood supply. The root, which has no enamel, extends down into the jaw bone and is held in place by a bone-like material called cement.

There are three types of tooth. The flat-topped molars are better at grinding, the front teeth or incisors are good for cutting, and the canines are used for gripping and tearing. There are only 8 molars, among the milk teeth, but 12 molars in an adult's dentition.

Sometimes the lower jaw is ahead and sometimes the upper jaw, but there is an approximate order in which the different types come through. The incisors are first. The forward pre-molars arrive before the canines, implying that grinding is more important than gripping, but the second pre-molars arrive after the canines. The teeth of girls appear before those of boys (on average), but boys lose their 20 milk teeth before girls. If there is rhyme or reason – in evolutionary terms – to tooth arrival, it is difficult to detect.

The milk teeth are lost in the same order of incisors, pre-molars and canines. The buds for the second set, the so-called permanent teeth, lie dormant until starting to grow in the sixth year. Once again girls are ahead, with their first molars (to the rear of the pre-molars) arriving at 70–72 months, a couple of months before the boys. The first of the permanent teeth therefore arrives before the first of the milk teeth has departed. The incisors are pushed out during the seventh or eighth year, and the canines in about the 12th year, with the pre-molars lost sometime in between. The second molars do not erupt until the 13th year. The third molars or wisdom teeth are even later, arriving during the 17th year, or perhaps not until the 25th year, or perhaps never. To accommodate the extra 12 molars the jaw bone must grow sufficiently. If it does not, the teeth become crowded and crooked. Many children need corrective treatment to ensure an orderly arrangement: a perfect set of teeth seems to be one of the few things that nature does not always provide unaided.

When the final molars are in place the individual concerned will have achieved the full complement of 32 teeth. Alas, no more will ever arrive. It would be convenient if another set erupted at, say, age 40. It might also be convenient if teeth continued to grow (as they do in rabbits), thereby replacing the worn upper layer. In humans the teeth have no capacity for growth or repair, leading to severe problems for many. Skeletons from ancient Egypt show that the teeth were badly damaged by sand in the food, and there is often evidence of what must have been agonising abscesses in the jaw. Today a diet rich in refined sugars provides plentiful food for bacteria, whose metabolism produces acid that attacks the enamel. Although teeth can rot during life, they survive longer than any other body part after death, their enamel being resistant to decomposition.

'hard wired' into the brain. In a study where such children were given glasses as toddlers, fewer than one-third had problems at the age of four. Their brains had become permanently adapted to receiving the 'correct' information. Children born with cataracts will be blind for life if the obstruction to the lens is not removed promptly.

The process of learning and rewiring is life-long, but by far the most rapid in infancy. In the first year, 60% of the child's food intake goes into providing energy for the burgeoning brain. The microscopic changes in the brain are mirrored on its surface, which gradually assumes the characteristic folds of the adult cerebral cortex.

There is no point in encouraging a baby to perform some task before the brain pathways are in place to make it possible. Outsiders, including parents, cannot know how an infant's nerve cells are being connected and disconnected, or how developed is the portion of the brain which controls some activity. Therefore they cannot know if the child is prepared and actually ready for that advance. No one expects a blind man to see if there is no optic nerve, but parents sometimes expect (or even demand) progress in their child's development before its time is due.

All the various capabilities take their time to emerge, and the times of their arrival vary between babies. However, on average, a baby can: lift its head momentarily at four weeks, readily at six weeks and recurrently at eight weeks; bear weight on its fore-arms at 12 weeks; raise itself to look directly forward at 16 weeks; shake a rattle at 16 weeks; grasp objects deliberately at 20 weeks; determinedly grasp its bottle at 24 weeks; suck biscuits enthusiastically at 32 weeks; sit without support at 32 weeks, and stand if supported; stand with the aid of furniture at 36 weeks; pull itself into a standing position at 40 weeks; lift one leg at 44 weeks; shuffle awkwardly at 48 weeks; walk, either with or without assistance, at 52 weeks.

There has probably never been a child who precisely matched this particular time-table, but these outward and visible signs of progress (indicating appropriate internal development) do all occur, more or less in that order, more or less at those times, and more or less without adult intervention.

Toilet training

Amazingly, human infants master the complexities of speech earlier than they achieve sphincter control, an ability that seems so simple by comparison. Awareness, and then control of urination and defecation, starts on average at 15 months. An early sign occurs when the child informs the carer that its pants are wet. By 18 months (so says Ronald Illingworth's happily optimistic textbook on *The Normal Child*) a child will be 'clean and dry with only the occasional accident'. By two years he or she will be 'dry at night if taken out in late evening' and, six months later, that child will be attending to its own toilet, 'save for wiping'. The infant is by then talking steadily, holding a pencil in its hand (rather than the fist), building a tower of eight cubes, helping to put things away(!), and giving its full name when asked for it. By age three, when parents are beginning to forget about excreta in either form, their child can copy a circle, know its own sex, ask questions (repeatedly), know some nursery rhymes, and may count up to ten – but there can still be 'accidents'.

Once again animals are superior, often depositing faeces in one location virtually as soon as they can walk. Humans, for all their intelligence, can be extremely backward, or merely negligent about activities readily accomplished by so-called lesser creatures. Watch a young human trying to put food inside its mouth. Watch it stagger (and trip) down the garden path. Watch it stare, wide-eyed with astonishment, as it pours forth some effluent, a sphincter having chosen to relax. Then watch an animal – and be humiliated.

What the human can do, and does whenever, however and wherever possible, is explore, touch, feel, try to eat almost anything, listen to sounds, push its finger everywhere, and generally gain experience of its environment. The human child is outstandingly inept as it stumbles, falls and trips, but its progress and curiosity seem deliberately ill-intentioned. It breaks things, pushes them over, creating mayhem as it moves – so indirectly, so accidentally – from A to B, or perhaps to F if that suddenly looks more interesting.

Yet that is its virtue. It is learning during every waking moment of every day. Its stubby, podgy fist will become the most dextrous of all limb endings.

GETTING MOVING

Achieving mobility is a compulsion for young humans. They *have* to move, to gain access, to explore beyond the restrictive sphere of their immediate surroundings. Most other animals rely on inherited, instinctive behaviour rather than the learned behaviour that predominates in humans. Think of an eagle flinging itself from some cliff-edge, having only flapped its wings beforehand. It does not crash, it flies. Humans, equally impelled to experience movement, crash and fail, again and again. Their progress is slower than that of any other animal. When trying to crawl (at, say, 36 weeks), they may actually progress backwards. The wish for movement is instinctive but not the style. There seem to be as many crawling techniques – on both elbows, on one, with one leg bent, with both, with neither – as there are babies. Some even omit this style of locomotion altogether, as if knowing that walking is the only proper way, however pathetically achieved at the outset.

Walking not only requires appropriate nerve connections and musculature, but also balance. Being bipedal instead of resting squarely on four legs means that balance is both harder to achieve and more important. A textbook reminds us of the complexity involved – 'A standing man can fall in any direction; forwards, backwards or sideways ... As soon as he begins to fall, reflex compensatory muscular reactions set in which restore the state of balance ... The muscular contraction then ceases till the next deviation from the erect position occurs.' All much easier said than done, as any one-year-old would tell us.

The organ of balance – the vestibular apparatus or labyrinth – exists within the inner ear, which consists of three distinct sensory systems. Firstly there are three semi-circular canals, each filled with fluid and each set in a different plane. All three act as a receptor system, with the contained fluid accelerating or decelerating, enabling the brain to detect and record head movement. Secondly there are the saccule and the utricle. These also work in a mechanical fashion. They possess chalky particles (otoliths, or ear dust) which respond to gravity (like 'snowstorm' toys which, if shaken, resemble snow falling on a village scene). The particles come to rest upon sensitive hairs, enabling the organ's owner to know which way is up and which is down.

These two systems permit balance to be achieved, whereas the third, the organ of Corti (see page 170), is the receptor for sound waves, measuring oscillations between 20 and 20,000 cycles per second. Not only does the vestibular system allow automatic reactions – we stand upright without giving this ability a second thought – but impulses from it also lead to the cerebral cortex, enabling us to think what we are doing and behave accordingly.

Below Progression from sitting to crawling to walking is achieved by practising each new skill again and again until it becomes automatic.

For the child learning to crawl and walk, nothing is automatic. Every movement must be learned through trial and error and practised until all its components can be committed to the brain and reproduced at will. Once again, the pathways of the brain are being moulded by experience. As the child manoeuvres itself into new positions and situations, its perceptual skills also develop. Even new crawlers hesitate at the top of a steep ramp, knowing that something is different about the terrain. But they are unable to adjust their movements to the information received through their eyes, and head recklessly down the slope, even when they have fallen several times. More experienced crawlers gradually adjust their movements to the angle, but when they began to walk the lesson has to be learned all over again.

On average, after first walking with assistance at 52 weeks, a child will: stand alone, temporarily, at 13 months; stand up without support, and struggle upstairs, at 15 months; run, and climb stairs by holding banisters, at 18 months; walk backwards at 21 months; climb stairs alone at two years; jump with both feet, walk on tiptoe at 2.5 years; stand on one foot momentarily, go upstairs one foot per step, and jump off the bottom step at three years; walk downstairs with alternating feet at four years; skip with alternating feet, stand on one foot for several seconds at five years; stand on each foot alternately, with eyes closed, at six years.

Many mammals would scoff at such leisurely progress, having run from their first days and leapt over obstacles very shortly afterwards. They would be further amazed that even six-year-old humans are still far from reaching maximum prowess. However those six-year-olds can do something utterly beyond the reach of even the cleverest animals: they can speak. And, just as importantly, they can understand when another human speaks to them.

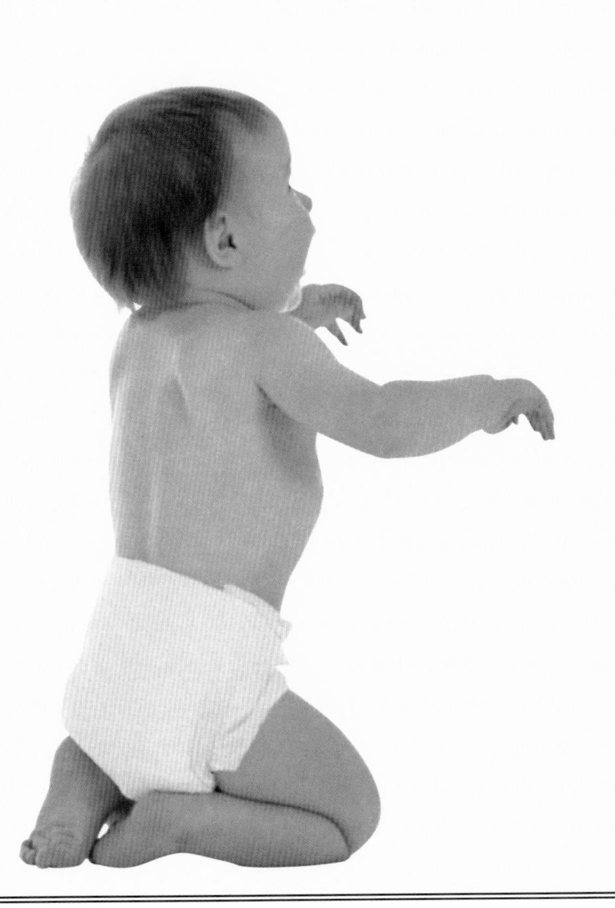

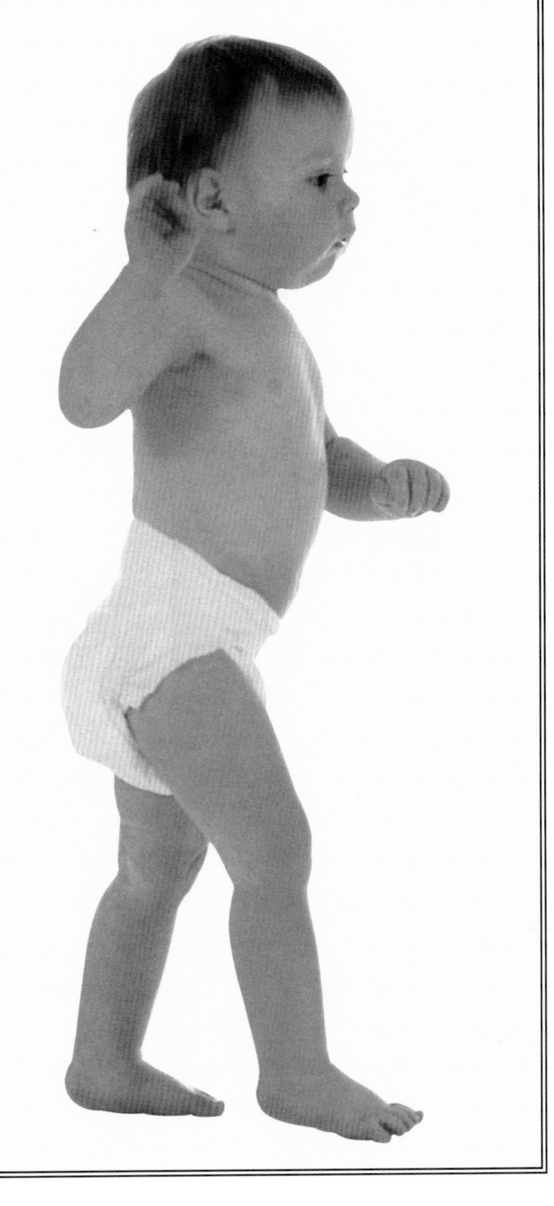

By using its eyes, its brain and its adaptable body in conjunction, the human infant will, when adult, become a full member of the most investigative and capable species the world has known. No less importantly it will have perfected its use of language, a brilliant means of communication that has shaped human achievement more than any other factor.

Language – the key to belonging

A newborn human has already heard muffled speech from inside the womb, but has not, until birth, made a sound. At the age of one it will say a few words. At two it will be saying sentences. At three it will jabber away enthusiastically, learning ten new words a day, or one every waking hour. At four it will almost certainly have mastered the complex abstract structure of the surrounding language – or even languages, if brought up in a multilingual environment. It applies the rules of grammar rigidly, speaking of 'drinked' and 'eated' rather than the irregularities of 'drunk' and 'eaten', and forming all plurals with an 's' to give 'sheeps' and 'deers'.

For the baby, acquiring language is a survival mechanism. The sooner it can speak, the more easily it can get its parents to do what it wants. Language offers a whole new way of expressing desires and feelings, and of manipulating others into supplying requirements that the child cannot yet obtain for itself. Equally, the parents can impart life-saving information, like 'don't eat the red berries on that tree, they are poisonous'. Beyond that, language is the key to entering the social world in which the child will have to exist for the rest of its life.

In the first month of life, crying is virtually the only sound made. In the next three months cooing, gurgling and laughing appear, along with specific cries with different meanings. By four to six months the infant is babbling steadily, making sounds that are universal no matter what the language of the surrounding culture. Even deaf babies babble. But despite this universality, babies at the babbling stage are homing in on the speech sounds being used around them. During the first six months they have a unique ability to make subtle distinctions between word sounds, even in languages they have never heard before. By the first birthday the brain has become wired in to the patterns of the mother tongue, and the infant is oblivious to the subtleties of foreign sounds just as an adult is. Nevertheless, the child retains a phenomenal language-learning ability for about the first six years. After this time, learning a language will never be as easy again, as schoolchildren will testify. Research suggests that people learning languages when older actually use a different part of the brain than the baby picking up language for the first time.

Understanding language comes before using it. As speech is a matter of mimicry, this has to be the case. It is probable that gestures are understood even earlier. Such actions would have been the principal means of communication before speech evolved, and can be extremely explicit. Some children learn and understand them so well that they are less concerned about learning speech, the gestures being sufficient to convey whatever it is they are trying to communicate. This situation is more likely to arise in families with several children.

Once children have grasped the concept of language, the speed with which they learn to talk is impressive, those ten new words a day quickly mounting up. A slowing down eventually occurs once the child has learned sufficient words for its needs. Two professors of paediatrics once 'very cautiously' offered a vocabulary development table, showing average words accumulated by certain ages. At 12 months the average total was three, rising to 22 at 18 months, 272 at two years, 896 at three years and 1540 at four years. The child of one, although starting from scratch, and not even knowing at birth that parental babble has significance, possesses a vocabulary of some 900 words by its third birthday. From then on, fully aware of language and exploiting it wholeheartedly, if not incessantly, the child adds another 1600 words in the next three years. Furthermore, many adults get by day-to-day with a vocabulary not much larger than that of a six-year-old.

As ever, there is great variation in language development. Speech comes later for twins than singletons, for boys than girls, for the hard-of-hearing over normal hearers, for sighted individuals over the blind, and for offspring with many siblings, particularly if several languages are being used around them.

At six to ten months an infant begins to imitate sounds, and show different responses to different words. First words are usually spoken between eight months and 2.5 years, appearing on average close to the first birthday. They consist of either single syllables or doubled up syllables such as 'mama' or 'car-car'. At 18–24 months the child learns to name a few objects correctly, and is pairing words together. Sentences and phrases begin, on average, by the second birthday, with a range between ten months and 3.5 years. The vocabulary is now growing fast, with nouns tending to be learned first, followed by verbs, adjectives, adverbs and pronouns. Hence 'cup', then 'give cup', 'give big cup', 'give big cup now', and 'give me big cup now'. The speech tends to be very repetitive. Infant talk has a characteristic sound, partly because the very young use more vowels than consonants while adults use one vowel to every 1.4 consonants.

By 24–30 months the vocabulary is shooting ahead to 400 words or so. Plurals, possessives and questions are being used. The average sentence is three words long. By 30–36 months the vocabulary may be a thousand words. The child often speaks extremely loudly. By its fourth birthday it is probably forming grammatically correct sentences. Stuttering is common, but no longer of whole words as it can be earlier ('give me cup-cup'). The voice is better modulated and controlled. By the fifth birthday grammatically correct sentences are almost certain to be used, and the vocabulary is around two thousand words. Sentences are about four words long, rising to five words at age six but not reaching seven words until about the age of nine. By the age of six the child can be considered to have mastered its local language and have sufficient vocabulary for its needs.

The child must not only learn words but how to say them. One reason why humans cannot usually learn another language perfectly after leaving childhood is because they cannot acquire its pronunciation. They find it difficult unlearning the kind of pronunciation they have developed (which is why, for example, many French people speak English in a recognisably French manner). Think even of two basic baby words – mama, dada – and they are created quite differently, with the first using the lips and the second the tongue plus the roof of the mouth. (Babies often say 'dada' first, presumably because it is easier.) They also watch, intently, as words are spoken to them, and this must help, but the blind also learn to talk.

There are five types of articulation in English, the first four being consonants. In plosives, or stops, the breath is halted to make them, as in 'pop'. In fricatives, or spirants, the breath is forced through a groove, as in 'thin' or 'fine'. For laterals the breath avoids the mouth's centre, as in 'low'. Trills involve vibration, as in the rolled 'r'. In vowels the air passage is more open than with consonants. There are also extras in other languages, such as the clicks of Xhosa (which English speakers usually only use to hurry on a horse). These configurations of lip, tongue, palate and teeth must all be learned early in the language process, and involve the coordination of over a hundred different muscle groups. Say any group of even monosyllabic words – pot, pan, tin, top, mum, give, push, food, dog, house – and remember these distinctions are all being skilfully acquired by humans still incapable in all sorts of other spheres.

The 50 commonest words make up 60% of those we say (even as adults) and 45% of those we write. But the relative rarities which make up the other 40% do total about two thousand, even for the simplest speakers. They all have to be learned, and infants mop them up with ease. They may still be stumbling, tripping, damaging, and turning the laden spoon sideways shortly before it finds the mouth, but they are acquiring communication with a different urgency. It is as if they know that the spoken (and the listened) word will be of greater value in the end. Or even straight away.

The theory of mind

Parallel with language acquisition comes the development of social skills. At the most basic level, the child must become aware that he or she is an individual, distinct from everyone and everything else. A one-year-old looking in the mirror does not see itself but another person. At 18 months the child will know

immediately who the mirror image is, and will delight in watching itself pull faces. The realisation is slower to dawn for identical twins, who are confused by the fact that the person in the mirror looks like their twin.

Crucial to social development is the understanding that other people have minds too, and do not necessarily share the child's own knowledge and desires. This is referred to by psychologists as 'mindreading'. As early as 14 months an infant will bring something to its parent specifically to show it to them, showing an awareness that the parent has an independent view of the world. Another sign that children are beginning to attribute independent desires and beliefs to others comes in play, when toys are given make-believe personalities and actions.

A classic test of more advanced 'mindreading' is the 'Sally-Anne' test. The child is shown two people, Sally and Anne. Sally hides a sweet under the middle one of three upturned cups, while Anne is watching. Anne then leaves the room, and Sally moves the sweet into the right-hand cup. When asked where the sweet is now, all children old enough to understand the question will usually reply correctly. But when asked where Anne will look for it, children under three or so will insist on the right-hand cup. They are unable to grasp that Anne can hold a false belief, that her knowledge is not necessarily the same as theirs. This test is measuring more than just logic or intelligence. Older children with Down's syndrome will expect Anne to look in the centre cup for the sweet, but those with autism, a condition characterised by an inability to relate to other people, usually fail.

The part of the brain responsible for 'mindreading' has recently been identified using magnetic resonance imaging. Scans taken while people performed

Children learn rapidly of their own needs, but it takes longer to appreciate that others have needs which may be different (*above*). Cooperative play is a sign that important social skills are being developed (*right*).

SPORT
CLUB

tasks using this skill show consistent activity in a distinct area. Presumably children are only able to develop a sophisticated awareness of the beliefs of others once this area is sufficiently well advanced.

There are many other mental leaps to make. The realisation that another child is suffering when it falls down and cries. The discovery that other people can be deceived through a lie. The misery of shame when the child knows it has done something wrong and that others will be displeased. The desire for acceptance, leading to shyness and embarrassment. Jealousy when a baby brother or sister receives attention. By the age of four, all the adult emotions are in place, and the child has a strong sense of self and individuality. Instead of merely owning a brain, it has developed a mind.

As all parents know, and learn more deeply as the years roll by, the growth of a child is far more than losing fontanels, imitating, controlling temperature, acquiring teeth, standing upright, balancing, vocalising, losing and acquiring reflexes, using hands, controlling sphincters, learning words, articulating properly, sleeping less, dreaming less, drawing circles, climbing stairs, changing cartilage for bone, getting bigger, stronger, cleverer, and generally worthier of being titled *Homo sapiens*. It is the development of a unique person, one who recognises its individuality, one who learns of truth (and lies), of shame and responsibility, one who knows that others have different thoughts and feelings – and that these other thoughts and wishes must co-exist with their own.

To quote Tolstoy once again: 'From the newborn baby to the child of five is an appalling distance.' From the child of five to adulthood is just 'a step', but there is one more hurdle along the way. Puberty, and the accompanying period of adolescence, is the rockiest stretch that remains on the path to adulthood.

● *Left* Childhood is a time of continual change – and yet lasts for 18 years. No animal takes as long as the human species to achieve adulthood.

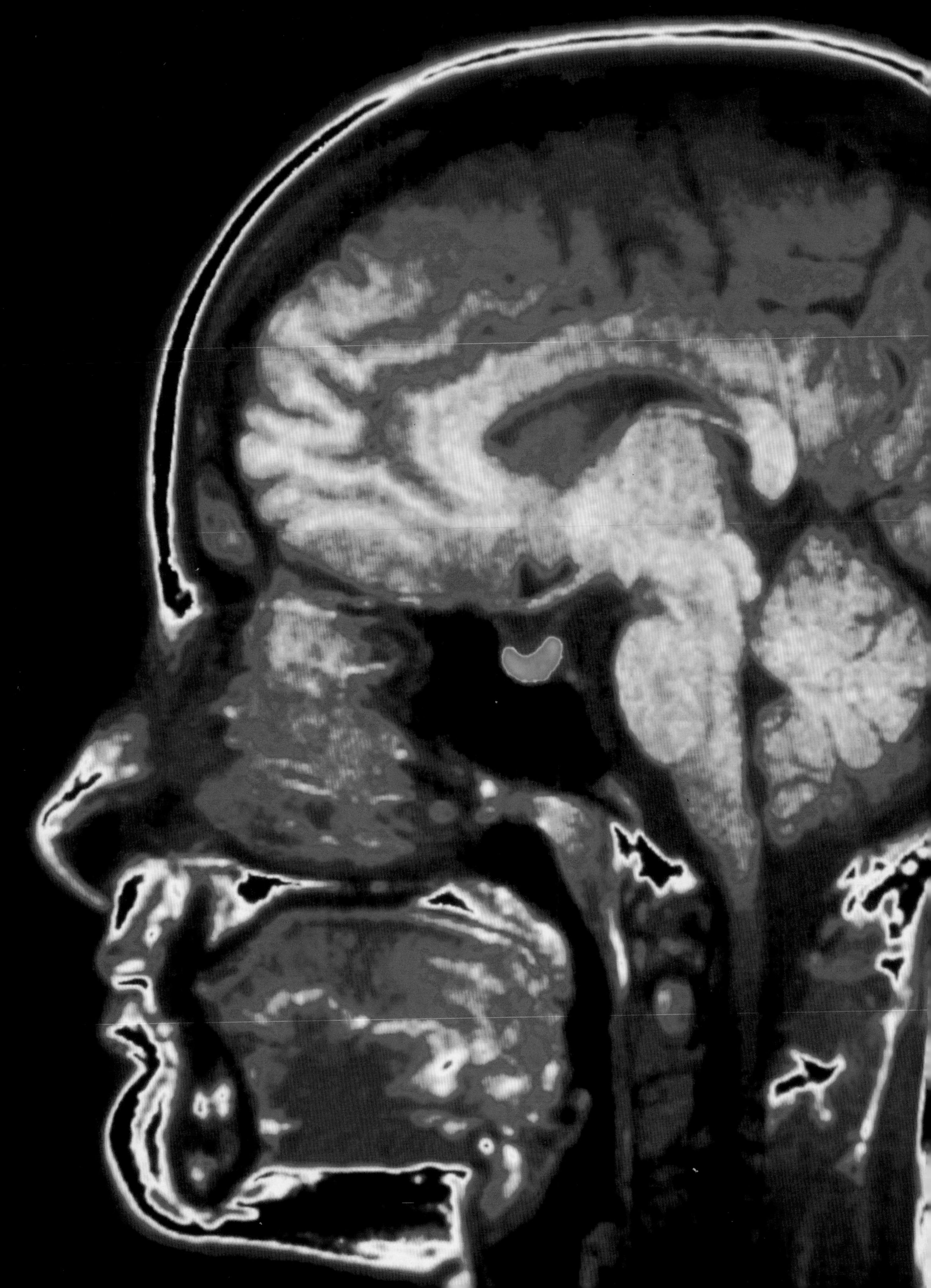

PUBERTY

We often have the illusion that somehow we are in control of our bodies, but, at least as far as physical events go, it is our biology that controls us. Rarely in life is this as obvious as during the great rollercoaster ride of puberty. We do not know how long it is going to take, or when it will start. And although we may think we know what is going to happen to us, nothing can prepare us for how it will feel.

Puberty is not only a time of confusion for adolescents, it is also something of a bewilderment to scientists. The precise causes of puberty, and why it happens when it does, are still a mystery. The nervous system and hormones are involved, but so too are social, psychological and nutritional factors. Pubertal changes are normally spread over four years, and, equally normally, can start anywhere between ages 8 and 13 for girls and 9 and 14 for boys. The process of puberty therefore embraces ages 8 to 18 within a normal range of children. As some children are abnormally early or exceptionally late, the pubertal span is even greater. This means that some may have completed their pubertal changes before others of the same age have even begun, causing considerable unhappiness to those who feel they are too early, too late, or just different. As girls start their changes 18 months or so before boys, girls may be taller, stronger and more sexually advanced than their male classmates. Once again this can be cause for emotional stress, perhaps misinterpreted by adults as one more sign of the bloody-mindedness of youth.

Left Sexual development relies on both genes and hormones. The pituitary gland, or hypophysis (*main picture*, green), called the master endocrine (hormone-producing) gland of the body, controls the onset of puberty. However, the X and Y chromosomes (the sex chromosomes) define an individual's gender (*inset*). The presence of a Y chromosome initiates the development of testes in the male foetus.

BEATRICE

Beatrice, of Salisbury in England, lives at home with her mother and her step-father. When she was aged 12 and still pre-pubertal, what changes were occurring in her body, and did she welcome them?

'I'm very aware of my height – and spots. My height is five feet one and-a-half. I'd like to change my nose, which is something out of hell, and my spots. I'm not dying for my periods, like some people are. I'm going to have a brace fitted for my back teeth. I don't have a boyfriend, and I don't want one. I've got other things to worry about, like my family and work. I'm not looking forward to my next birthday because then I'll be a teenager. You get breasts and all that, and I just don't want it to happen, really.'

A year after this first interview, Beatrice celebrated her 13th birthday and reviewed her previous 12 months. She did not sense much change in herself, and was not moody. Running was painful because bras were 'really, really uncomfortable'. Her armpits had grown hair – 'ugh, I have to shave now'. She had become taller, but still had 'nothing on the hips'. She did not want her periods to start because a friend 'has got them and she gets really, really moody'.

Growing up, in Beatrice's opinion, had little to be said for it. 'Some people want to grow up. I don't like the physical side. I'd rather it all appeared one morning, after I'd been in a cocoon for a while.'

Beatrice was still 13 when the most blatant sign of female puberty actually arrived. 'It caught me completely off guard. I don't feel like a woman at all. But ... in a way ... I'm pleased my periods have started. The good part is that I know I'm a girl, a female. The bad part is that I know I'll go on having periods. In a way I can't wait until I get my menopause!'

Triggers for change

Like most things in the body, the onset of puberty is orchestrated by the brain. The hypothalamus, a small structure in the centre of the brain about the size of a grape, begins secreting large amounts of 'gonadotrophin releasing hormones'. These in turn stimulate the pituitary gland, also in the brain, to produce two hormones known as LH and FSH. At first they are released at night, coming in regular pulses as the child sleeps. As puberty progresses their secretion is stepped up, until it reaches the adult pattern that will continue throughout life.

LH and FSH are identical in males and females – it is their effects on their target tissues that vary. In males, LH attaches to specific cell groups in the testes and stimulates them to increase their production of testosterone.

Interestingly, testosterone production also depends on the hormone prolactin, better known in females for stimulating lactation. This most male of hormones is the key to sexual maturation, enlarging the testicles and developing their sperm-producing structures, enlarging the penis, stimulating hair growth on the face, chest and pubic area, increasing muscle mass and bone density, expanding the larynx or voice box, and awakening the sex drive. FSH binds to another type of testicular cell and regulates the production of sperm.

In females, the pituitary secretes its hormones in a monthly pattern. FSH, whose full name is follicle stimulating hormone, causes egg follicles to ripen in the ovary. Even before ovulation begins it stimulates the ovaries to produce oestrogen, the key to puberty in girls. Oestrogen causes fat to be laid down on the breasts, buttocks and thighs. It has a special effect on the pelvis, making the bones flatten and widen. Oestrogen receptors in the skin make it soft and smooth, whereas the male skin reacts to testosterone by becoming thicker and tougher. LH controls levels of that other female hormone, progesterone, which plays a crucial role in the menstrual cycle.

These chains of events are well understood. The mystery is what triggers the hypothalamus to initiate the sequence in the first place. What triggers the trigger? Researchers suspect another area of the brain is involved, perhaps stimulated in its turn by feedback from the body.

Whatever sparks it off, puberty is definitely happening earlier these days, at least in the developed world. Average age of puberty has advanced by some 18 months since the beginning of the twentieth century. The age of menarche (first menstrual period) in the UK was 15 in 1890, and is now 13. However, the trend seems to have stopped or slowed in the last 20 years or so. There has been much speculation about why puberty has got earlier. The fact that the modern world is brighter than in the past has led to the suggestion that changing light levels, so important as a seasonal sexual trigger for many animals, are also affecting humans. However, most researchers think artificial light is not strong enough to be so influential.

Nutrition has definitely improved, and is generally believed to play a role. Average weight at the start of puberty for girls is still 47 kilograms (103 lbs), just as it was four generations ago. The equivalent male weight is unchanged too at 55 kg. The difference is that better nutrition means that the critical weight is reached at an earlier age. Is this the cause of earlier puberty, or is the weight issue coincidental? The enigma remains.

There are also environmental influences. The timing of puberty has been described as an interaction of geographic, ethnic and genetic factors with socioeconomic status, health, nutrition and emotion. Moderately obese girls tend to start earlier. Those with a fear of obesity, or a fear of growing up, can be later. Anorexia nervosa delays puberty, as can stress, excessive exercise and inadequate diet. But the commonest cause of extreme pubertal delay is physiological abnormality, usually genetic.

Puberty's arrival is not so much the start of sexual maturity as its re-start. Considerable sexual development occurs within the womb, its pattern set by the sex chromosomes. Males have an X chromosome and a Y chromosome, the X supplied from the mother and the Y from the father. Females receive an X from both parents. Possession of a Y chromosome causes the gonadal (reproductive) tissue to begin forming testes at six weeks post-conception. Prior to this the gonadal tissue is identical in both sexes. If there is no Y chromosome, the embryonic gonads develop into ovaries instead. At seven to eight weeks the developing testes start producing the masculine hormone testosterone, which is essential for the further development of the male sexual organs. In the absence of testosterone the foetus will develop female genitalia, regardless of its chromosomal sex. Similarly, a genetically female foetus exposed to high levels of testosterone would develop male sexual characteristics.

The dual nature of sexual development, with its reliance both on genes and hormones, occasionally results in gender ambiguity. This happens when the hormonal environment in the developing foetus contradicts the chromosomal blueprint. In the most common case, a genetically male baby lacks receptors for testosterone. Although the hormone is produced normally the cells are unable to recognise it, and so develop to a female pattern. Externally the genitals appear female, and the child is therefore raised as a girl. Internally the vagina ends blindly and there are no uterus or Fallopian tubes. Problems only come to light at puberty when the girl fails to menstruate. It is also possible for female babies to be confused with males because an excess of testosterone has caused genital enlargement. Sometimes the genitals are ambiguous, in which case genetic tests must be used to determine the child's gender, followed by surgery if appropriate.

The brain also experiences sexual differentiation during the foetal period. Circulating hormones not only affect its components (male and female brains show differences in patterns of neural connections, the size of cell nuclei and aspects of metabolism) but also influence tendencies towards certain personality traits and abilities. Despite the enthusiasm for treating boys and girls in the same way, they are not born the same (quite apart from their distinctive physical differences). No traits can be rigidly classified as either male or female, and personality and ability are influenced by many other factors besides gender. However, behaviour and social responses are influenced by sex differences, as most parents of young children will testify.

Following birth, very little change occurs to infant reproductive systems; there is an 'arrest of maturation' that is unique to humans. For most animals, the journey from newborn to fully grown adult happens as a continuous process. Ewe sheep show the first evidence of sexual maturation at six to seven months, pigs at seven months and cattle at 12 months. Mice reach a similar stage in just 30–35 days. Gorillas, perhaps a fairer comparison, show signs of puberty at eight to nine years, markedly earlier than the human average. If humans followed the typical animal pattern we would be able to reproduce at four and would have stopped growing by the age of six. Instead, our bodies actively switch off sexual development. The process of reaching sexual maturity simply as a consequence of getting older is stopped, and instead we wait nearly a decade before our bodies start preparing themselves for the vital function of reproduction.

The timing of sexual maturity in a species, like the stage of development at which its young are born, is a trade off between different evolutionary pressures. For most animals, it would be evolutionary suicide to spend 14 years in the

Crystals of human sex hormones. Testosterone, the prime male sex hormone, is shown above and oestradiol, the most potent of the six oestrogens (female sex hormones), is on the left. These hormones control sexual development and the secondary sexual characteristics, such as facial hair and breasts.

dangerous wild environment without getting down to the crucial business of passing on the genes. But our human ancestors were different. They needed the extra time for something so vital to their survival that even sex could wait. That something was learning – learning to function in the increasingly complex societies humans were creating. Until the youngster could hold its own in the social group, reproduction was out of the question.

Only with puberty does the second era of hormonally controlled development take place. This second stage further increases sexual differentiation, building on the foundations laid before birth. The role of the foetal stage of sexual development tends to be underestimated, but it is thought that hormonal events before birth influence a child's response to the stimuli that control the onset and the effects of puberty.

It has been suggested that pheromones, chemical messengers released by the body for communication purposes, may play a role in triggering puberty. Pheromonal communication is best known among insects; for example, female moths lure males from several kilometres away by releasing pheromonal sex attractants. In humans, the role of pheromones is controversial. Some scientists are sceptical of their very existence, and others believe that our species has lost the ability to detect or respond to them. However, pheromones could explain some interesting examples of what appears to be sexual communication between humans. For instance, women who live together tend (after three or four months) to synchronise their menstrual cycles more than would be expected by chance. Male pheromones have been implicated in the timing of ovulation and the length of the menstrual cycle: the timing of sexual maturity in female mice is substantially earlier if they are exposed to males than if they are kept in isolation.

Right Adolescence can be a time of confusion, of wishing for independence but maintaining dependence, and of being attracted to the opposite sex but (often) spending more time with the same sex.

Adolescence

The arrival of sexual marking points – menarche for girls and first sperm for boys – will occur on a definite date, but the influences leading to these events are numerous and, so far, not fully understood. As for adolescence, it cannot be said to begin on any single date for anyone. It is a period of transition and not a moment of attainment.

Adolescence has long been regarded, notably by adults, as a time of turmoil and disturbance, of 'storm and stress', according to one writer in 1904. Children of those days, leaving school early, starting work immediately and never known as 'teenagers' (with money and time to spare) are nowadays assumed to have

SPOTS

To countless adolescents, acne seems like sheer bloody-mindedness on the part of nature. Just when they are especially concerned about their facial appearance and how it might appeal to others, along comes a plague of spots precisely where it matters most.

Eighty per cent of teenagers suffer from spots to some degree. Problems range from a few occasional pimples to widespread and painful inflammation. The face is the most common area affected, but acne can also appear on the neck, upper back, shoulders and chest – the areas of the body that have the largest sebaceous glands.

The sebaceous glands secrete an oily substance called sebum, and are at their most active in adolescence. When an excess of sebum blocks up the pore belonging to a hair follicle, a spot is formed. Boys are more prone to acne than girls because sebaceous secretion is stimulated by testosterone and inhibited by oestrogen. Girls often find that spots flare up just before their period, when oestrogen levels are at their lowest.

For some reason that is not known, the walls of the hair follicles in acne sufferers become thickened, and thus block more easily with sebum. The blocked follicles are known as comedones. Bacteria trapped in the comedones begin to multiply and produce toxins that cause the skin to become red and inflamed, and the walls of the follicles begin to thin as the pressure inside them builds up. Eventually the follicles rupture. If the swelling is in the upper layers of the skin a white pustule appears on the skin surface. If it occurs deeper down it forms a nodule or cyst, a tender swelling lacking a white head. Squeezing these nodules spreads the inflammatory products further into the skin and makes the situation worse. Sometimes they form the characteristically pitted acne scar.

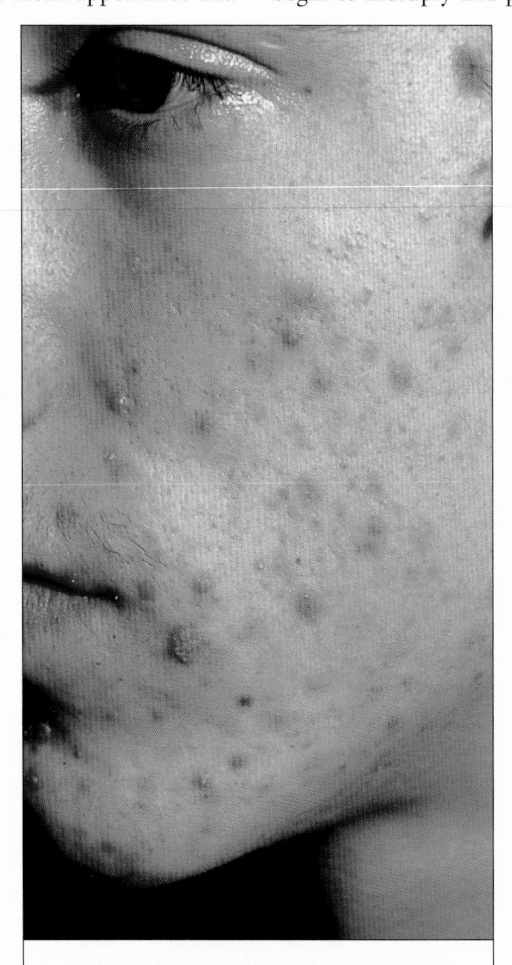

● *Above* Acne. The oil-producing sebaceous glands are more likely to become blocked during adolescence. The plugged ducts are then susceptible to secondary infection producing inflamed pustules.

● *Left* Blackheads appear when hair follicles become plugged with dried sebum.

Some comedones form without inflammation. If the top of the follicle dilates and exposes the sebum to the air the result is a blackhead. If it remains closed it forms the white raised nodule under the skin known as a white comedo – particularly common on the forehead.

Many myths have grown up around acne. Lack of washing, lack of exercise, eating too much fat and sugar, 'hormone imbalances' and even sexual activity are all commonly believed to contribute, but none of them is guilty. Eating too many chips won't cause acne, but working in a fast food restaurant in a hot and greasy atmosphere might. Although hormones play a part, levels are normal in the great majority of sufferers. Washing has a limited effect, as the spots are caused by oil from within the skin rather than surface dirt. Stress can make acne worse.

The good news for today's teenagers is that acne can be successfully treated in a variety of ways. Luckily too, the curse is only temporary. Only 5–10% of young adults report problems with acne.

been obedient, respectful and knowing their place. The writer's remark is therefore surprising, save that each new community of adults seems to repeat it. The youthful realisation that parents can be wrong, less than god-like or merely normal is often called delinquency or insufferable by those same parents, but is hardly reason for despair. The traditional view of the teenage years is mistaken, according to many current child psychiatrists. As one phrased it: 'Most young people pass through adolescence without developing any form of significant mental disturbance.'

Psychiatric disorders are commoner in adolescence than in middle childhood, but the rates are thought to be similar to those in adult life. Before the age of 12 suicide is rare, committed by fewer than one in ten million. From 12 to 16 there is a hundred-fold increase, and another ten-fold rise from 16 to 19. This thousand-fold leap is huge and disturbing, and could be taken as a violent indictment of the adolescent period, save that suicide rates actually continue to rise with age and reach a maximum in the elderly. Teenagers are therefore behaving as young adults rather than aberrant children. Attempted suicide is, similarly, extremely rare in childhood and far commoner in adolescence, especially its later years. Actual suicide is commoner in boys/men, but attempted suicide (parasuicide) is more frequent in girls/women.

Depression also takes a leap in frequency when children reach their teens. Animal phobias, common in the young, do not increase with adolescence, but social phobias and agoraphobia do. Anorexia nervosa emerges, notably in girls between 14 and 17, but can also occur in the pre-pubertal years and in boys. Abuse of drugs becomes commoner with adolescence, reaching a maximum at 18–21. Male adolescents and young men have the highest rates of criminal offending of any group in society. But according to one report, 'socially disruptive behaviour, such as rebelliousness, cruelty and aggression, does not alter greatly in rate over the 8- to 15-year age period.' Therefore, to generalise, delinquent children are likely to become delinquent adolescents, and a hard core who begin crime in adolescence will carry it on as adults. 'Storm and stress', so popularly linked with adolescence, are therefore not the fault of that phase, but either a hangover from childhood or the first stage of adult life.

Puberty is not the same as adolescence, but occurs as part of it. Adolescent features such as becoming taller, heavier, stronger and cleverer are matters of degree. Puberty is quite different, with sexual prowess shifting from inability to competence, from infertility to fertility, and thus to the ability to pass genes on to a further generation.

Modern Western society cares greatly, almost exclusively, about chronological age. This can conflict with sexual maturation, and can make a nonsense of progress towards adulthood. More traditional communities have tended to regard pubertal changes as more significant, caring little or nothing for actual age.

A total of 186 cultures were once examined in a major survey, and almost all were found to embrace rituals linked to puberty. Such rites of passage were commoner for girls than boys, with the girl ceremonies usually linked to menarche and rituals for boys taking place a year or more after sexual maturity. Female rituals tended to be family or local group affairs, and male equivalents involved either local groups or larger gatherings. Boys tended to be initiated as a team, and girls more individually. Actual rites varied greatly, with boys generally performing tasks, experiencing hardships, learning about 'group cohesion' and, less frequently, suffering mutilation. Girls were taught more about taboos and their particular roles within the community. Genital surgery was commoner for males than females, and is still more frequent for males in modern societies that maintain such rituals. However, the genital mutilation still undergone by girls in many societies, sometimes in infancy rather than at puberty, is generally far more brutal and damaging.

The survey found almost as many variations on pubertal rites as there were cultures, despite some common trends. Rituals in the developed world have been reduced to buying a first bra or tampon, or, on the male side, even less. As for sexual intercourse for adolescents, the communities examined held views at both extremes. One-fifth of the communities looked at both expected and approved of pre-marital sex. Three-quarters did not consider female pre-marital sex 'a particularly serious issue', but one-quarter thoroughly disapproved. Males were twice as likely to be initiators of any pre-marital sex, but the decision was usually mutual.

As well as preparing the body for sex, puberty also prepares the mind. Sex hormones in the blood actually cross the blood-brain barrier into the brain and connect with waiting receptors, just as they do in the body. Suddenly, sexual arousal begins to occur. The opposite sex (or the same one) are transformed from objects of disdain or ridicule into fascinating, desirable beings. Teenagers begin to experiment with physical and emotional relationships, culminating, for many, in sexual intercourse.

The American author Mary Hotvedt has described different societies' reactions should an unmarried 15-year-old girl become pregnant. 'If she is a rural Iraqi peasant, her father may kill her. If she is a Kalahari bush-woman, she may marry quickly and comfortably. If she is a middle-class US high school student, she might stay unmarried and attend a special class for expectant mothers.' The same student would be more likely to see the event as a disaster and have the pregnancy terminated. Teenage pregnancy is rarely welcomed, partly because it expresses the contradictory nature of adolescence. Very young mothers are simultaneously showing independence (in having a child) while remaining dependent on others (their parents or social welfare provisions).

In Western society, sex tends not to wait for the legal watershed of the age of majority; indeed, the age of sexual consent is usually lower than the age at which a person is considered adult, and the age of sexual maturity is often lower again than the age of consent. This emphasises the biological fact that sexual maturity comes earlier than other aspects of maturity. About 17% of American girls have had an abortion or have given birth before reaching 18. Twenty per cent have had an abortion before the age of 20 and a further 20% have given birth. If natural miscarriages are added to these figures it is probable that half of all girls in the US have conceived before leaving their teens. As intercourse can only lead to pregnancy on some 60 days a year, and as contraception is so readily available, the proportion of adolescents involved in sex is assumed to be far greater.

Another American survey found that the mean age for first intercourse in girls was 16.9 years, with the range spread from 10 to 27. Eighty per cent of girls had had sexual experience by the time they reached the age of majority at 18.

A British survey in 1990 found that one in four adolescents, both boys and girls, had had intercourse before the age of consent (16). The average first-time age was 17. A sample of 55–59 year-olds questioned at the same time had not lost their virginity until an average age of 21.

Male development

Contrary to popular belief, a boy's testicles do not 'drop' at puberty but descend while he is still in the womb. However, puberty marks the beginning of sperm production controlled by the pituitary hormones, LH and FSH (see pages 86 and 87 for more on this subject).

Sperm are produced in the seminiferous tubules of the testicle. This process was examined in more detail in chapter two, 'The Beginning', which looks at the reproductive process.

Sexual maturation also brings the ability to achieve erections. As further evidence that sexual development occurs in two stages – before birth and at puberty – many male infants have erections and fondle themselves. Without erection, the mere possession of a penis does not enable the male to inseminate. Humans are the only primates without a bone, the os penis, to assist in stiffening. Instead, distension and rigidity are caused entirely by raised blood pressure within the penis, which is rich in cavernous spaces that fill with blood when the arteries leading to them dilate. The invasion of blood and subsequent swelling also blocks the veins which normally drain the area. This engorgement increases the volume of the penis and makes it rigid, a state known as tumescence. During detumescence, or relaxation, the arteries constrict, permitting the veins to enlarge and the accumulated blood to flow away. Females also

Right The human testis. On the right are the seminiferous tubules where the sperm are created before being moved along to the epididymis (top), a convoluted tube where they complete their development and are stored ready for ejaculation. At ejaculation the sperm pass into the vas deferens (left), and on into the penis.

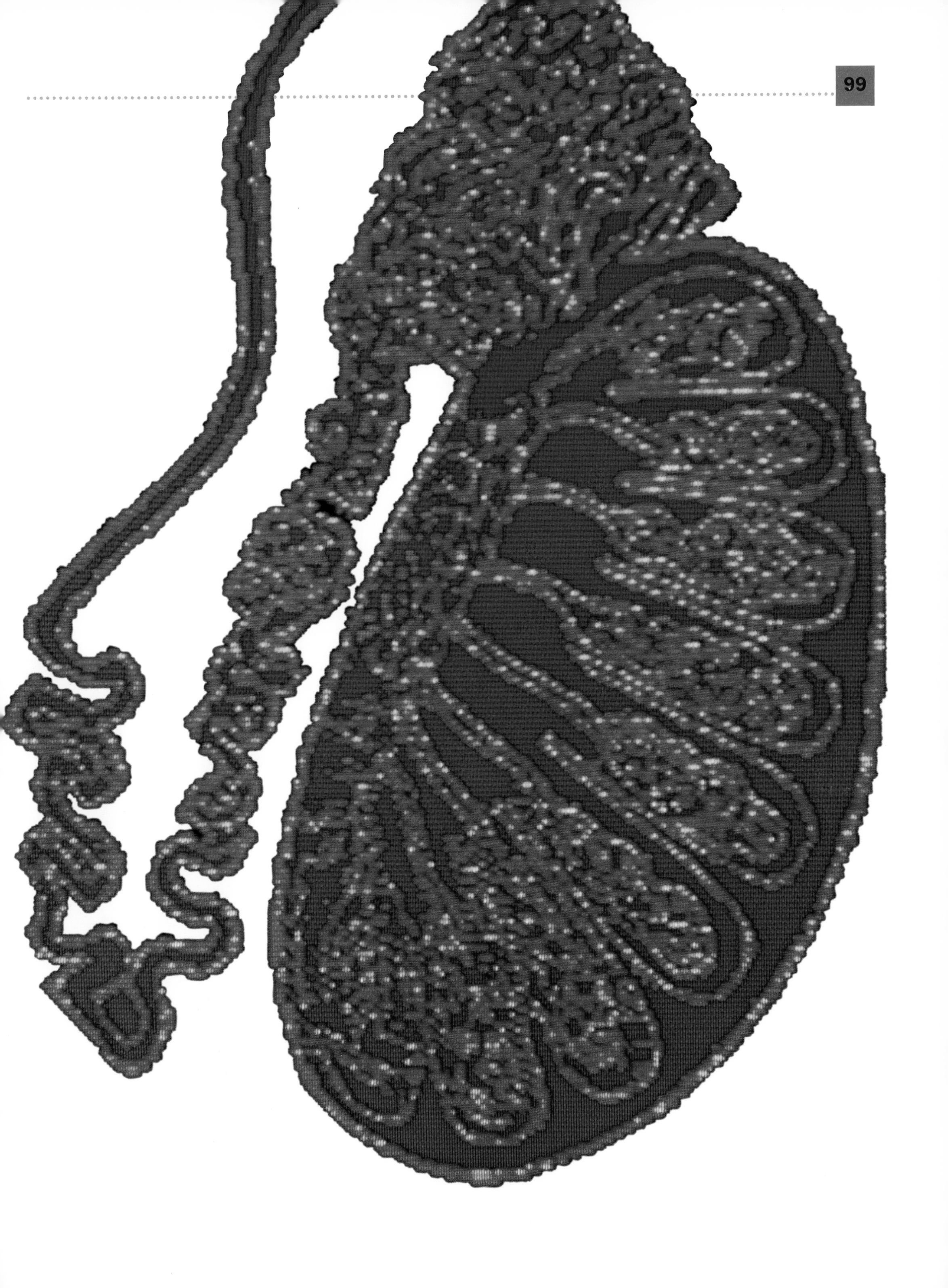

HAIR, SMELLS AND SWEATING

The growth of hair in new areas is another important sign of puberty. There are hairs all over the body even in prepubescent children, but these are fine and relatively inconspicuous. At puberty the hairs on the legs and arms become more pronounced and new growth appears in the pubic area, under the arms, and, for males, on the face and chest.

● Beard hair grows slightly faster than scalp hair, causing a daily chore for those who wish to be 'clean-shaven'. Its growth is about 0.38 mm a day, and the so-called five o'clock shadow therefore represents 0.14 mm of new hair since shaving time. The picture (*right*) shows shaved beard hair seen through an electron microscope.

Pubic hair, although forming different configurations in men and women, is similar in structure in both sexes. It is also similar across all ethnic groups, despite extreme dissimilarity in head hair (from straight to tightly curled) and considerable variation in facial hair (from virtually none in Amer-Indians to vigorously bushy in Nordic men). Pubic hair is always short, coarse and curly. It is oval in cross-section, whereas straight hair is round. It has a shorter growth period than scalp hair, this lasting only about six months before the follicle shuts down and the hair falls out. The hair therefore has no time to grow to a longer length.

Adolescents may wonder why all this new hair appears, and so do scientists. One suggestion is that pubic and underarm hair provide a greater surface area for the dissemination of odours from the scent glands in these regions. Our individual smells are unique (ask any dog!) and it would seem that pubertal hair growth has survived despite modern humans' lack of interest in smell in general, relative to most mammals. The real oddity is that odours that begin during puberty, which – presumably – are linked to intercourse and reproduction, should cause such discouragement in modern humans rather than attraction. The role of odour has not only diminished in importance but seems to have switched from good to bad. We still find smells attractive in the opposite sex, but we tend to prefer those from perfume factories to those emanating from the armpits or genitalia.

Linked to the production of adult odours at puberty is a change in the pattern of sweating. Humans have two types of sweat glands, each with very different functions. The eccrine glands are found all over the body and produce a faintly acid, watery fluid containing less than 2% solids. The evaporation of this fluid from the skin helps to keep the body cool. This process is continuous, but sweating only becomes noticeable when secretion levels rise in response to hot conditions or exercise. Humans rely on sweating to regulate temperature far more than most animals (for example, we do not pant) and are capable of losing huge amounts of fluid in this way. Eccrine sweat is odourless, only beginning to smell when broken down by bacteria on the skin surface.

In contrast, the apocrine glands secrete a thick, waxy, greyish substance that is diluted by fluid from other glands nearby. In most primates, including the big apes, apocrine glands exist all over the body. They develop similarly in the human foetus, but after the fifth month they disappear from all but a few areas: the armpits, navel, ears, nipples and pubic zone. These glands do not become fully functional until puberty. They have no role in temperature regulation but are related to the scent glands of animals, and are thought to function solely for odour secretion. They are not stimulated by heat but by stress and emotion, including sexual excitement. Once again, the initial secretion is odourless, at least to the modern human nose. Its strong scent follows soon afterwards as bacteria set to work upon its fatty acid and protein constituents. Consequently it is the smell of decomposition rather than the smell of the substance itself that we now find repugnant. In other primates, apocrine odour from the armpits is thought to act as an erotic attractant. Plainly something has gone wrong or is different among humans, either with their sense of smell or with their armpits.

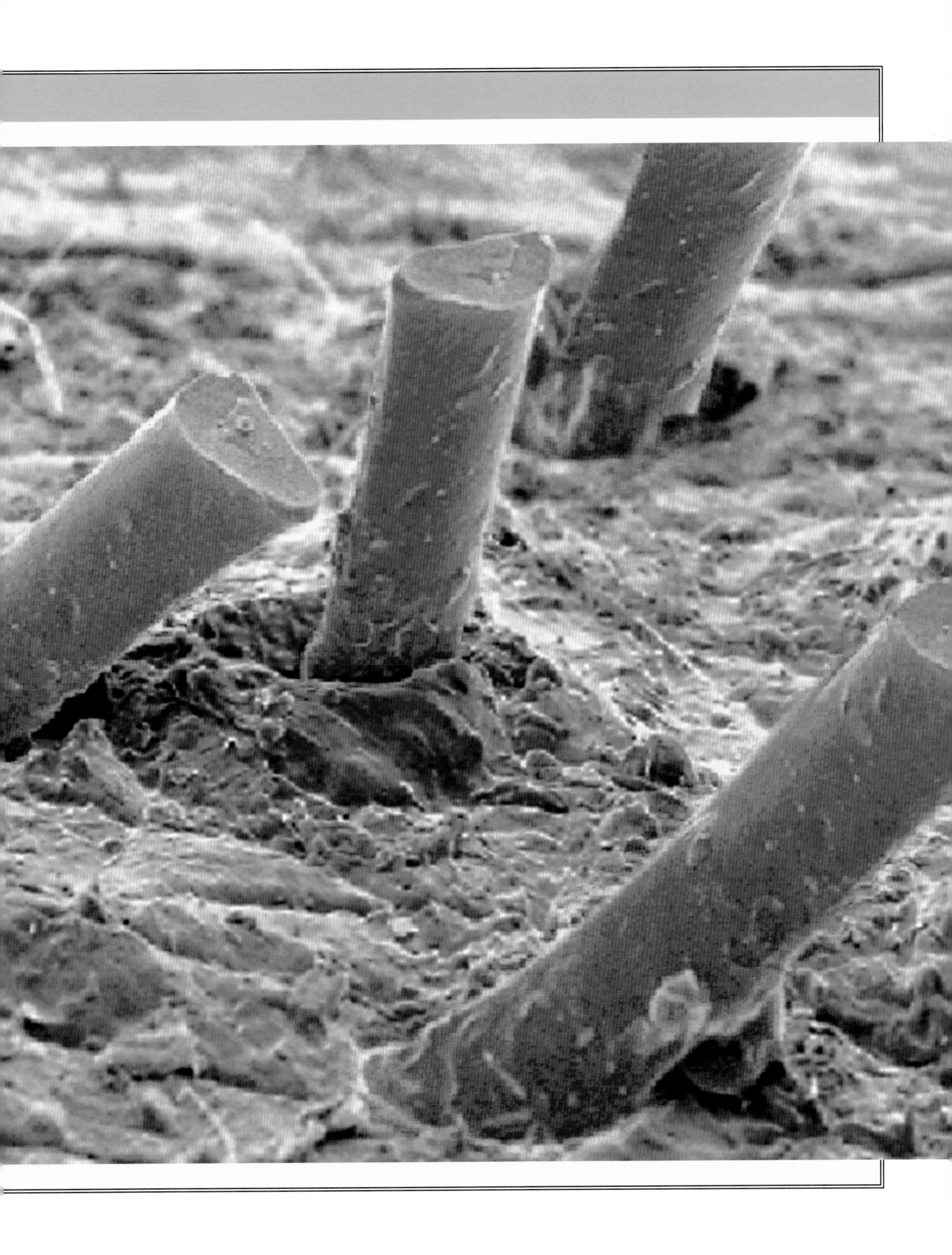

possess erectile tissue, in the clitoris and around the mouth of the vagina. This swells in response to sexual stimulation and relaxes when the stimulus is removed.

Voices change for both sexes at puberty, but most markedly in boys. Males of many species have a deep, booming voice to attract females and intimidate their rivals, and the human voice possibly evolved to meet similar requirements.

The male larynx or voice box starts its growth spurt when the penis is completing its period of accelerated development. The laryngeal cartilage is very sensitive to testosterone, and on receiving the hormonal signal it grows larger and heavier. The vocal cords, which are membranous folds lying across the airway or trachea, grow longer and thicker to give the deep male voice. The larynx sits at an altered angle, so that it projects conspicuously to form the so-called Adam's apple. Girls with an excess of testosterone may also develop such an apple, but normal laryngeal growth is such that a girl's voice drops only a couple of tones in puberty, compared with the octave drop for boys. Voice change is usually gradual, extending over several years, but boys can sometimes croak haphazardly when their voice is 'breaking'. Nothing is actually being fractured, but uncertain muscular control over the growing larynx can cause the warbling. About half of boys start their voice change at 15 but, like everything connected with puberty, its timing is extremely variable.

A rough chronology of male puberty is given below. All the age ranges given are averages, and perfectly normal development may occur on either side of them.

- The earliest sign of puberty is growth of the **testes**, starting anywhere between ages 9.5 and 15. Half of boys show testicular enlargement by their 12th birthday and 97% by their 14th. The brain stimulates the testes to produce the hormone testosterone in increasing amounts, causing the sperm-producing structures to develop.
- The **penis** subsequently starts to enlarge, as do the prostate gland and seminal vesicles, both of which secrete substances that are added to the ejaculate. There is rapid growth of the genitals between about 12 and 15.

- **Sperm production** then begins, often in the 13th year but with a normal range from 12 to 16. First ejaculation takes place about one year after the penis begins its enlargement, either through masturbation or during sleep (a 'wet dream'). Sperm become mature around age 14–16.

- **Pubic hair** usually starts to grow before the growth spurt has begun, first appearing at 12–14. It goes through five stages: resembling the hair on the rest of the abdomen; slightly pigmented and still downy; definitely pigmented and curly round the base of the penis; curly but less abundant than with adults; adult extent. It is abundant by the age of 13–16.

- **Armpit** and **facial hair** do not grow until an average of two years after the first showing of pubic hair. Average age of appearance for armpit hair is 14.3 years, achieving adult distribution by 15.6 years. Facial hair appears at the corners of the upper lip at about 15, and on the cheek at 16. It does not usually spread to the chin until genital growth is complete.

- The **sweat glands** in the armpits and around the genitals increase in number and activity.

- The **growth spurt** (in height) does not begin until the sexual changes have been initiated. In general, the sooner puberty begins the sooner the rate of growth will decline and finally stop. Half of boys will have reached their peak growth velocity by their 14th birthday and 97% by their 16th.

- The larynx grows and the **voice** begins to break when the penile growth spurt is drawing to a close.

- The **skin** alters its texture and appearance.The skin of the scrotum darkens.

- The **breasts** may enlarge a little at around 12–14 before returning to normal at 14–17. The ring of tissue around each nipple (areola) enlarges.

- Almost all boys experience **acne**.

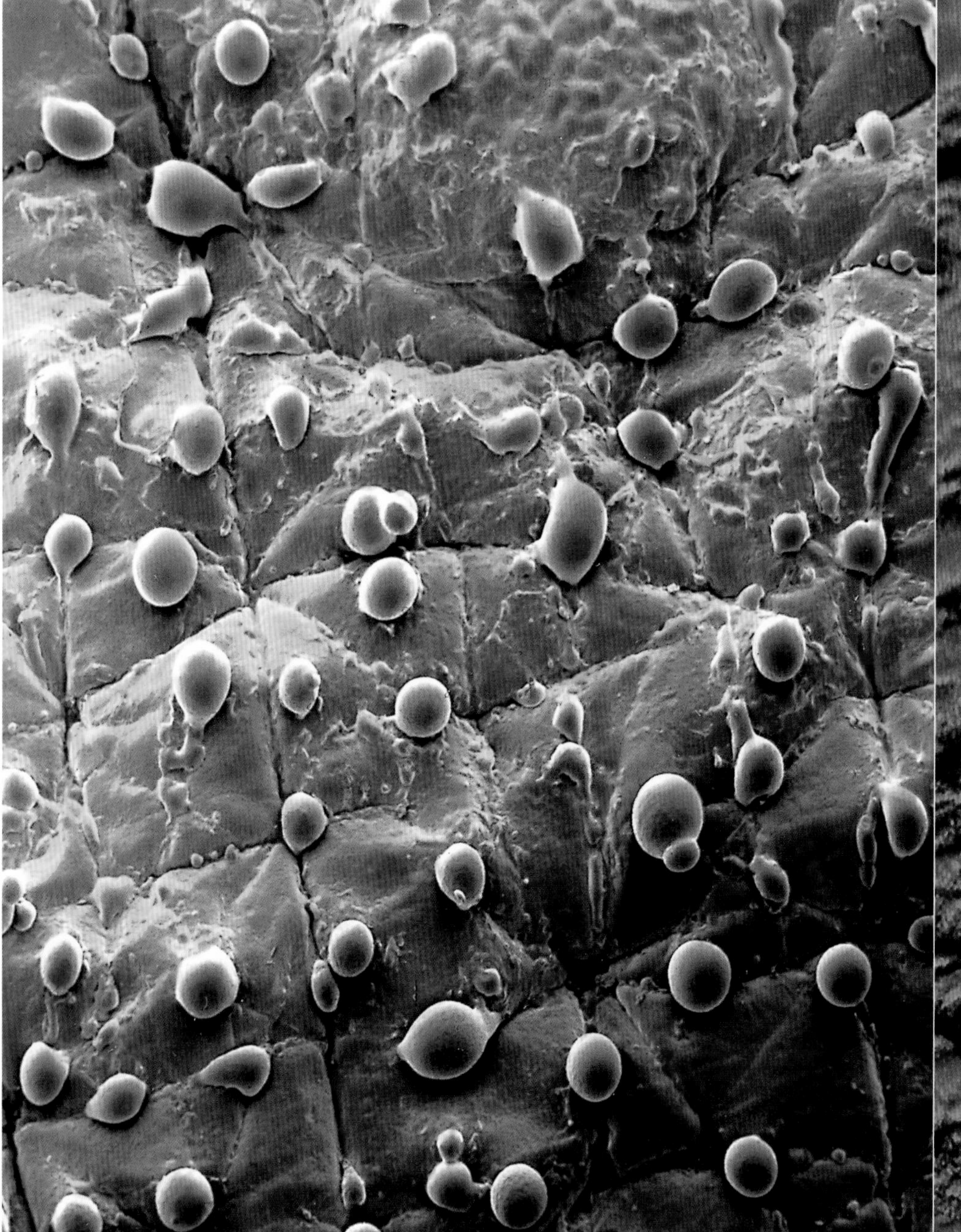

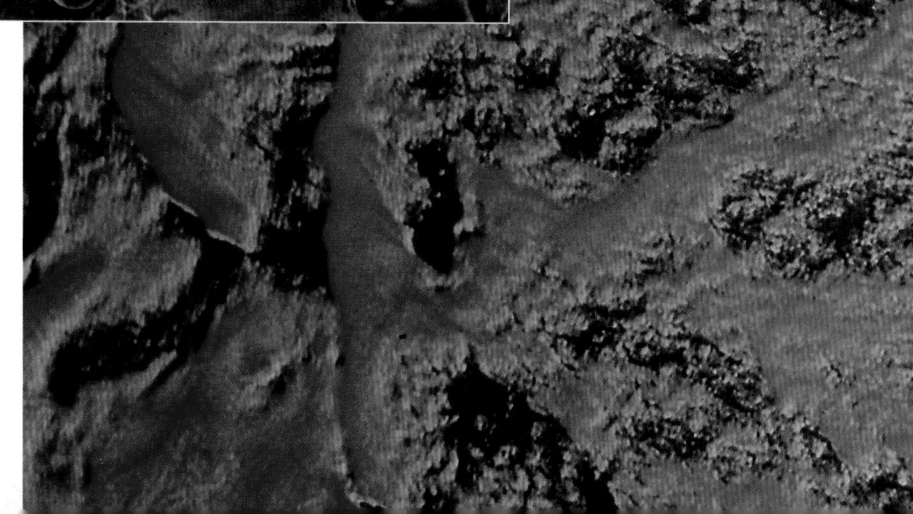

Above Sweat droplets (in blue) being produced from the back of a hand after exercise. Humans are the only species with sweat glands all over their bodies, sweating being our principal cooling device.

Right Electron microscope image of a sweat pore. Sweat is secreted by sweat glands within the skin and emerges through the pores.

Female development

Breast development, the most outwardly conspicuous sign of puberty in girls, undergoes its initial stages before birth. It begins at about six weeks post-conception with a thickening of the skin on the so-called milk or mammary line. This exists in both males and females and stretches on each side of the foetal body all the way from the armpits to the groin. The thickening then dies back and only persists in the eventual location of each breast. Babies are sometimes born with signs of other breasts, usually only one and generally rudimentary, a condition known as polythelia. Such extras always occur somewhere along the mammary line and can appear in either sex.

By birth there are several columns of cells, each leading inward from the breast bud. Both sexes possess such a system, which may even produce a little secretion shortly after birth (so-called 'witches milk', maternal hormones having acted as stimulant). Nothing else happens in the breasts of either sex until puberty arrives.

At the onset of puberty the rising levels of oestrogen and progesterone in a girl's blood cause fat to be deposited around the breast in a characteristic pattern. The duct system develops from the primitive arrangement laid down before birth, but only partially. Some secretory glands are formed but the breasts do not become capable of producing milk until pregnancy, when the huge surge in hormones causes further growth in the ducts and a proliferation in the number of glands. Many girls and women also notice swelling and tenderness in their breasts in the second half of the menstrual cycle. This is due to the action of female hormones, which increase the amount of water retained in the tissues.

Women who have given birth in the past are able to produce milk again if there is a need, even after a lengthy gap. However, only around half of those who try this relactation eventually succeed, and they rarely produce more than half the required amount. The breasts tend to sag and flatten with age as the stimulating effect of oestrogen diminishes.

Male breasts are exactly the same as female ones during the pre-pubertal years. At puberty there may be a temporary lag in male hormone production, and boys may begin developing breast tissue (gynaecomastia) before androgens (male hormones) swamp the background oestrogens. Small amounts of liquid may be secreted. Up to one-third of boys experience some breast development on one or both sides. The tissue will regress within one to two years when production of the male hormones increases, but needless to say the condition can cause great worry and embarrassment for the affected boy. Adult men who take female hormones will develop breasts. In elderly males, whose production of testosterone diminishes, oestrogens may once again become dominant. Most men over 70 therefore develop a quantity of breast tissue.

There are stories that men have lactated in response to suckling, but the substance they produce, if the tales are true, is not expected to be milk.

Below is a summary of the events and timings of female puberty. Again, the age ranges mentioned are all averages, and it is quite normal for events to happen earlier or later. Interestingly, puberty tends to begin earlier if there are several older brothers and sisters.

- An increase in the size of the **ovaries** (invisible) is the first sign of puberty. Oestrogen in the urine provides proof of ovarian activity.
- Enlargement of the **breasts** is the first external sign, occurring in distinct stages. First the nipple rises above breast level, then the areolae rise and enlarge. Fat is deposited and the breasts fill out, initially in a conical shape. Finally each areola sinks level with the rest of the breast, which takes on its rounded adult shape. Development may start as early as nine. Half of girls will show signs of breast development soon after their 11th birthday, and 97% soon after their 13th birthday. The breasts are fully mature at 16–18.
- The **uterus** enlarges and the **vagina** accelerates its growth. Vaginal secretion begins at 11–14.

THE MENSTRUAL CYCLE

For girls, by far the most significant event of puberty is the onset of periods, known as menstruation from the Greek word for month. Periods are the outward manifestation of the menstrual cycle, the rhythmic pattern of events that controls female fertility. At its root is the monthly ripening and release of an egg from the ovary and the preparation of the uterus to receive an embryo.

Ovulation is caused by a surge in the production of luteinising hormone (LH) by the pituitary gland in the brain. Levels of LH and follicle stimulating hormone (FSH) as described on pages 86 and 87 begin rising from the first day of menstruation, which is defined as day 1 of the cycle. FSH causes a number of egg follicles to begin ripening, and in turn the ovary steps up its secretion of the hormone oestrogen. Two days before ovulation the pituitary steps up its hormone output, and the ovary responds by switching some of its secretion from oestrogen to progesterone. At the same time, the most mature of the ripening egg follicles ruptures to release its ovum, while the rest of them die off. Ovulation occurs on day 13–17 of the cycle, depending on the individual.

The empty ovarian follicle is then transformed under the influence of LH into the corpus luteum (yellow body), which produces increasing amounts of progesterone until day 22, when LH from the pituitary starts to decline. Unless the egg has been fertilised the corpus luteum then shrivels. A fertilised egg begins to produce human chorionic gonadotropin (HCG, the hormone picked up in pregnancy tests) and this takes over from LH in maintaining the corpus luteum for the duration of the pregnancy. The corpus luteum sits on the ovary and reaches about 1.5 cm in diameter.

All through the first part of the cycle, oestrogen has been causing the lining of the uterus to thicken. At the beginning of the cycle it is less than 1 millimetre in depth, but by ovulation it has reached 3–5 mm. In the second half of the cycle, when progesterone dominates, it becomes thicker still. Its blood supply increases and its cavities become distended with mucus, ready to receive a fertilised egg. When or if hormone secretions from the corpus luteum drop, the blood vessels in the lining constrict for a short time and the tissues die. Haemorrhages then develop as the vessels open up again. The presence of the discarded material makes the uterus contract, sometimes painfully, and its contents are expelled over several days. Even as this is happening, the pituitary gland is beginning the cycle all over again.

Blood loss averages about 40 ml (2 fluid ounces), but can be five times more or five times less. The length of the cycle also varies between individuals, averaging 28 days but ranging anywhere between 19 and 37. Some women have no regular pattern and go for long spells between periods. For young girls, the first few cycles may occur without ovulation until the hormonal pattern is fully established.

Pre-menstrual syndrome is experienced to some degree by as many as 75% of women. Breast pain and water retention are common symptoms and are likely to be the result of changes in hormone concentrations during this part of the menstrual cycle. Some women report irritability, tiredness, nausea or migraine, and accidents in the home are more common. There may be a craving for chocolate or stodgy food. Occasionally there is severe mental disturbance: much female crime is said to be committed during the pre-menstrual phase.

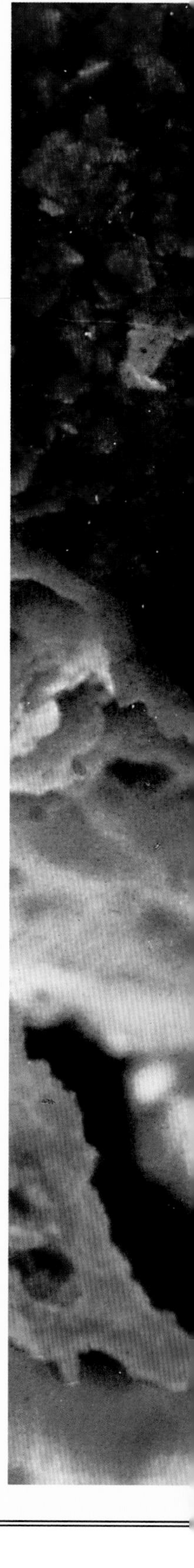

Ovulation. A fluid-filled follicle (*left*, pale pink), containing a single egg, on the surface of the ovary. At ovulation the follicle ruptures (*right*), and the egg is expelled. After floating across a small gap it enters the Fallopian tube or oviduct.

- The **pelvis** alters its proportions, and female contours start to appear as fat is deposited. Girls start to accumulate fat around the hips from the age of eight or ten.
- **Pubic hair** develops in stages as with boys, but ends in different configuration. It first appears at 10–12 and is abundant and curly by 11–15.
- **Armpit hair** appears at 12–14, the average being 12.5 years. It has an adult configuration shortly before 14.
- The number of **sweat glands** in the underarm and genital areas increases.
- **Menarche** (first menstruation) occurs on average 2.5 years after the start of breast enlargement, and normally between ages 10 and 16. Half of girls will have started menstruation by their 13th birthday, and 97% by their 15th. Malnutrition, extreme exercise and emotional stress can delay menarche, but most variation between individuals is thought to be genetic. Menarche of twins comes later than with singletons. Girls may not be sexually mature (i.e. producing fertile eggs) for 1–2 years after menarche. On the other hand, some girls produce eggs in a regular pattern even before their first menstruation.
- The **growth spurt** (in height) is generally finished before menarche; girls are unlikely to grow more than 7.5 cm after their first menstruation.Tall girls achieve sexual maturity before short girls, but girls with late puberty grow – on average – taller in the end.
- Onset of **acne** is very variable – if it does appear it is usually between 12 and 16.
- The larynx (and therefore the **voice**) changes, but less markedly than with boys.

The growth spurt

A sudden increase in height, known as the growth spurt, is a major component of puberty for both sexes. Girls start their spurt an average of two years earlier than boys. Peak growth rates occur, on average, at 12–13 years in girls and 14–15 years in boys. Growth is almost completely finished by the ages of 18 and 20 respectively. At the peak of their growth boys will gain about 9 cm per year, a centimetre more than girls, but these very high growth rates last only a few months. Peak growth starts at the bottom of the body and works upwards, so teenagers are likely to outgrow their shoes first, followed by their trousers and then finally their jackets. Most of the height increase comes from the trunk, but because the legs grow first and most rapidly, leg growth seems more noticeable. It is well known that people have steadily become taller in the last couple of centuries, but the change in 'sitting height' (the length of the trunk) has been far less than that in full height. Almost all the historical height increase has been due to increased leg length.

The pubertal growth spurt accounts for about 20% of final adult height, and it also brings about the final and permanent cessation of growth. Growth hormone, released by the pituitary gland in the brain, stimulates the deposition of protein, and therefore growth, in all parts of the body. A deficiency of growth hormone results in the condition known as pituitary dwarfism, where the body proportions are normal but stature is very short. However, growth hormone alone does not control growth. Growth hormone is produced by the body throughout life, levels in the adolescent being similar to those in adults. It is the body's response to the hormone that alters.

Increase in height is brought about by extension of the long bones, which grow from special areas of cartilage at the ends. First new cartilage is deposited, and this is then converted to bone. During puberty the increased levels of sex hormones cause the growing areas of the cartilage to be closed off, so that by late adolescence the growth cartilage has all been used up. The ends of the bones then fuse and can never grow again, no matter what stimulation is given.

Above A group of 12 youngsters aged 13. Not only does the growth spurt occur earlier in girls than boys, but there is a very wide range in its age of onset. The later the spurt occurs, the taller the individual is likely to end up.

Right A bone cell in a human femur. The cell (centre) is surrounded by the bone matrix. The cells are continually produced and resorbed so that the entire skeleton of an individual is replaced over seven years.

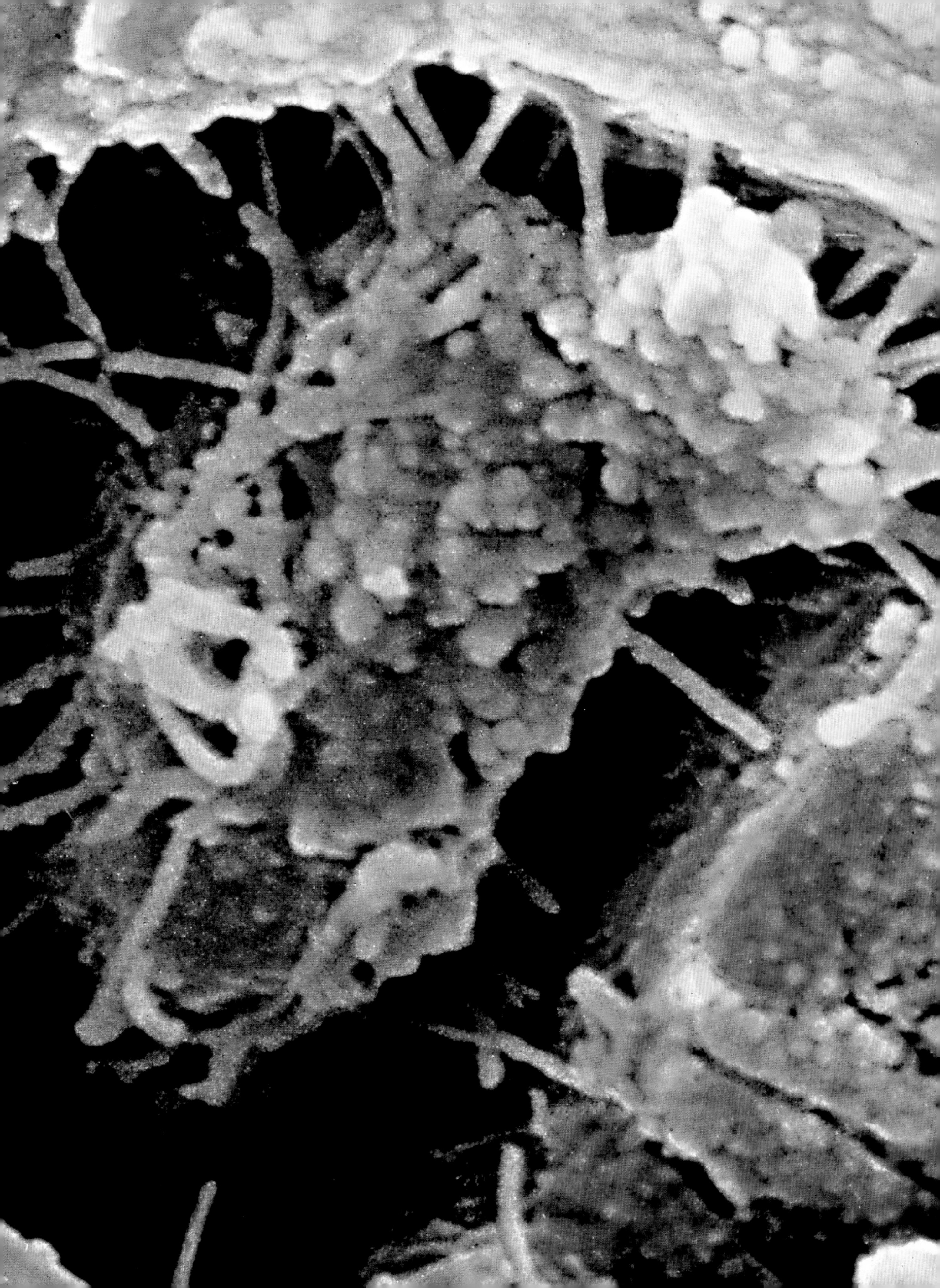

Men are generally taller than women for three reasons. Firstly, boys are slightly taller than girls throughout the pre-pubertal years. Secondly, boys start their growth spurt on average two years later, giving them two additional years of pre-pubertal growth before the final surge. Lastly, their peak growth velocity during puberty is greater, especially in the trunk. The average gain in sitting height during puberty is 15.2 cm for boys and 13.5 cm for girls, but leg length grows an average of 12 cm and 11.5 cm respectively, a much smaller differential. The net result of all these factors is that men end up an average of 12.5 cm or 5 inches taller than women.

The final height reached does of course vary tremendously between individuals, with 'normal' adults spanning a range of at least a foot (30 cm) within each sex. Equally, some women are taller than some men. Today, the most important factor influencing adult height is genetic make-up: tall parents are likely to have tall children, though the variation between siblings can be marked. Twins teach us much about the 'nature versus nurture' or heredity versus environment debate in many areas, and height is no exception. Non-identical twins, who share the same environment but have different genes, can differ considerably in height. Identical twins show the same growth patterns and reach very similar final heights. However, the influence of genes can be overridden. In the past it is likely that poor nutrition and frequent disease prevented people from reaching the full height of which they were genetically capable.

Height is not the only aspect of the growth spurt. All the bones grow, and their density increases, the latter more so for males than for females. The facial bone structure changes dramatically, with the nose and jaw becoming more prominent. The body takes on the characteristic male or female contours. In girls the hips become wider to facilitate childbearing. Their pelvic bones grow at the same rate as those of boys, but because the rest of the body grows more slowly the hips become proportionately larger. In boys the shoulders and rib cage become broader and the arms proportionately longer.

Right A dexa scan showing the fat distribution in a human female. The darker areas just under the skin are fat and the lighter grey areas next to the bones are muscle.

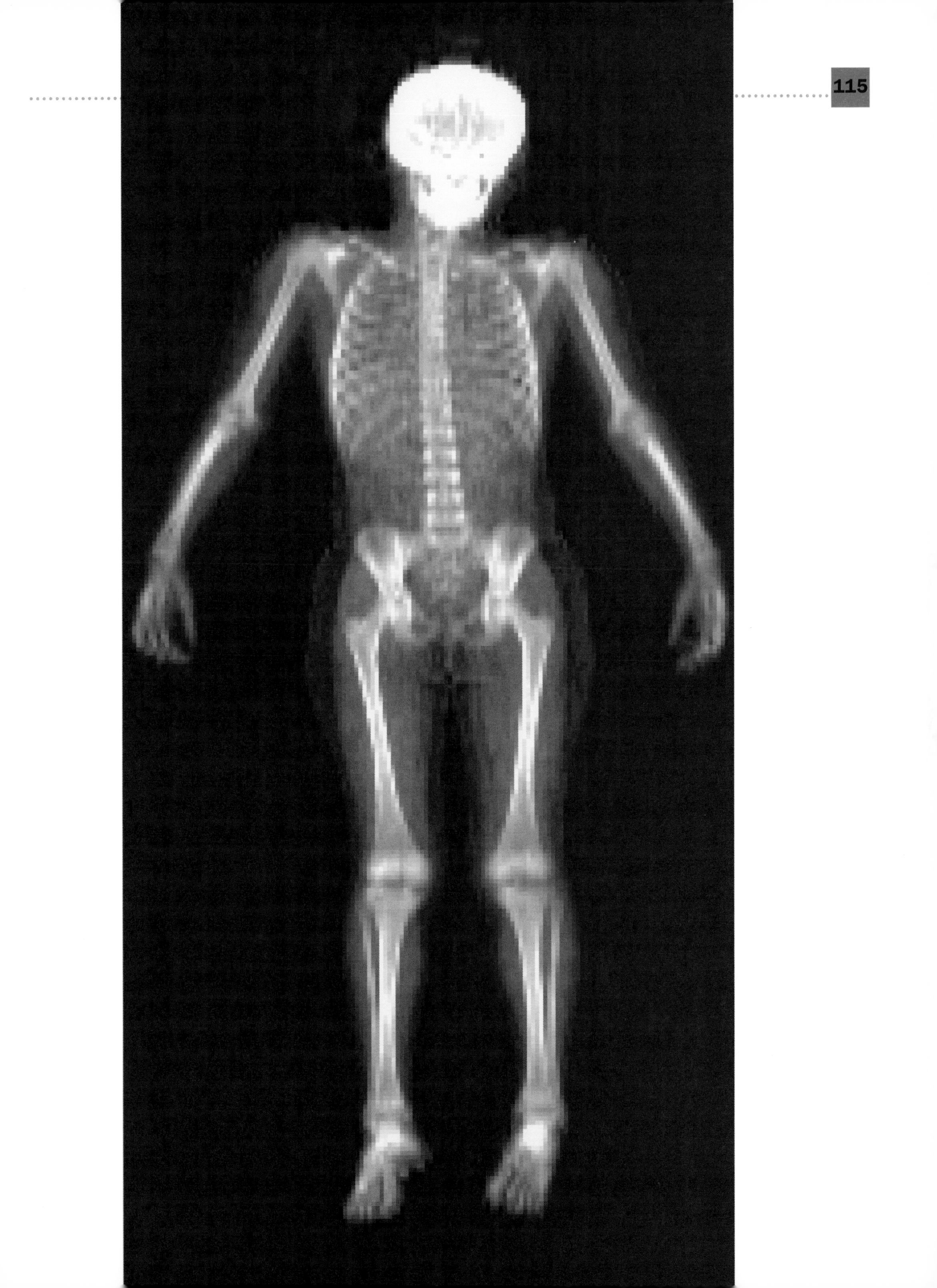

Fat distribution also changes. The lean body mass and the proportion of fat are almost identical in young children of both sexes, but changes start at six for girls and nine for boys. During their adolescent spurt boys may actually lose some fat, especially on the limbs. By late puberty fat accounts for about 12% of male body weight. In girls fat is deposited continuously, the limbs typically becoming fatter during the time of maximum growth. After the height spurt they lay down fat more rapidly and extensively than males, particularly on the breasts, buttocks and thighs to produce the curves of the female form. The ratio of lean body mass to fat changes from 5:1 before puberty to 3:1 at menarche. Fat composition is thought to play a role in starting and maintaining the menstrual cycle: thin girls begin menstruating later, and the periods often stop if fat falls below about 17% of a woman's body weight, for example in intensive sports training or anorexia.

As everyone knows, particularly those whose task is to keep kitchen cupboards stocked with food, puberty can be a time of huge appetite, notably for boys. In fact, teenagers' energy, protein and water requirements are all less per unit body weight than those of younger children. A baby in its first year needs 110 kilocalories and 4 grams of protein per day for every kilogram of body weight. A teenager between 13 and 15 needs only 60 kcal and 1.5 grams of protein per kilogram per day, and in adulthood the requirement drops to 40 kcal and 1 gram of protein.

These figures do not mean that babies eat more than adolescents – far from it. For example, males aged 13–15 will weigh perhaps 60 kg (or 20 times more than a baby). Therefore these teenagers require 60 x 60 calories (3600) per day, as against 110 x 3 (330) for babies. Similarly, the same adolescents require 90 grams of protein per day as against 7 grams for a baby. Even though the teenagers are 20 times heavier than the babies their energy requirements are only 11 times greater and their protein requirements 13 times greater. Babies, after all, double their weight in their first year. Adolescents, despite their growth spurt and their relentless discarding of outgrown clothes and shoes, never approach a doubling of weight in any year (for which the cupboard stockers should be most grateful).

Physical and sexual changes are not the only aspects of puberty. During the adolescent years the whole way that we think is transformed from child-like to adult. In childhood the brain is like a sponge, soaking up new experiences and information, learning language, mastering control of the body. During puberty the goalposts shift. There is evolutionary pressure to become independent of our parents and to reproduce, and a child's mind is no longer good enough. To match our adult bodies we must develop minds capable of analysing and reacting to the complexities of adult society. We carry on learning from experience throughout life (except perhaps towards the very end), but that ability to become wiser is laid down in adolescence.

When we enter puberty we are children, perfectly adapted up to a point, but unable to fulfil the biological imperative of reproduction. By the time we emerge from it we have crested the last ridge and are gazing at the summit that all our previous development has prepared us for. We still have much to learn, but biologically we have arrived. Physically, mentally and socially we can hold our own as independent members of the human race. And, if we choose, we can pass the genes that have formed us on to the next generation.

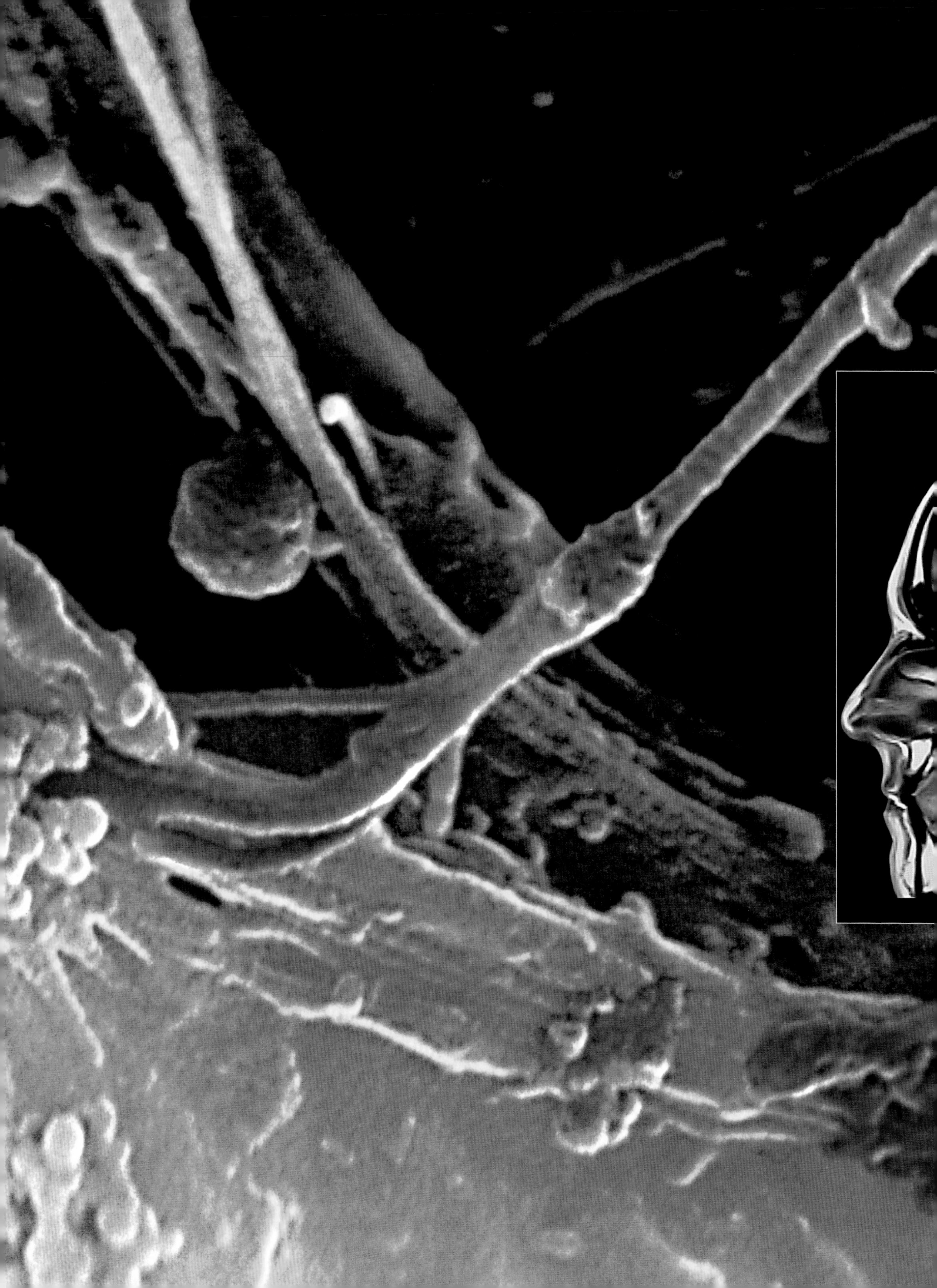

ADULTHOOD

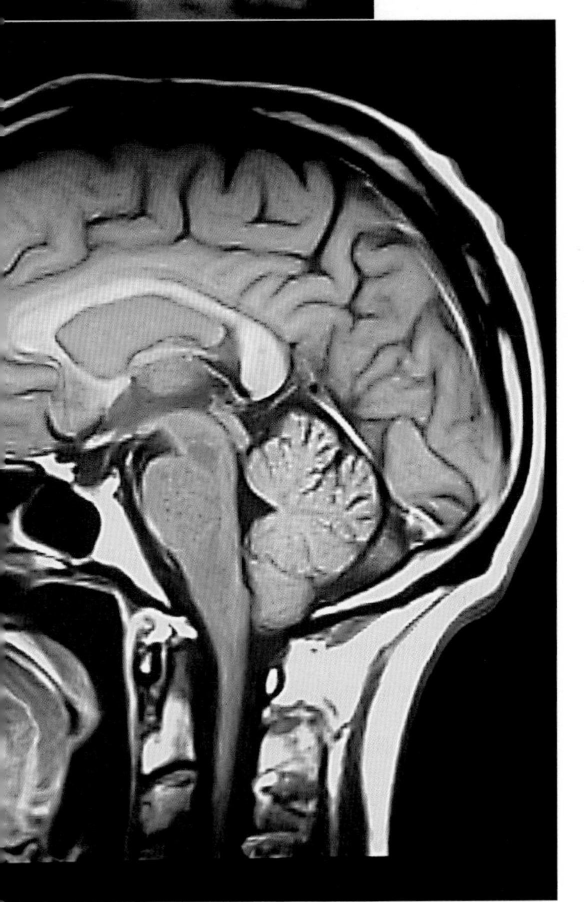

Conception is a hasty business, the liberated sperm and egg needing to meet before their brief survival time is up. Pregnancy is also rapid, with a single cell becoming billions in just nine months. In infancy and early childhood, mental and physical progress can be measured almost by the day. The pace of change then slows down a little until puberty arrives, bringing with it the transition from child to adult human being. Finally, after something like 20 years, the purpose of all that change is accomplished. Another example of Homo sapiens, *dextrous, brainy, talkative, creative and able to reproduce its kind, is standing firmly on its own two feet.*

As a species it has done extremely well. It has invaded every kind of habitat – mountains, desert, frozen wastes, forest, tundra, savanna, island, swamp. In the past, with isolated communities and poor communication between neighbouring groups, people could not know if they alone inhabited the earth or if other humans existed over the ocean, beyond the forest, across the mountain range. We now know, all too vividly, of the 5.8 billion people sharing our planet. Human beings inhabit a greater proportion of the Earth's land surface than any other

Main picture Nerve fibres (purple) forming connections with a nerve cell body. Chemicals released from the swollen tips of the fibres pass on signals from their parent cells. These signals are responsible for relaying information to and from the brain, governing the way we think and behave.

Inset An MRI scan of the human head, showing (in blue) the spinal cord leading to the brain, the cerebellum (in green) which controls movement, and the tremendous bulk of the cerebral hemispheres (brown), the location of thought and intelligence.

MARSHA

Marsha is 47 and is a mission specialist – an astronaut – with NASA. Her first space flight was an 11–day mission on the Space Shuttle Columbia. Since then she has left the earth another three times, most recently in 1997 to dock with the Russian space station, Mir.

On earth, Marsha is by any standards a highly skilled individual, equally at home solving complex engineering problems or at the controls of a large transport plane. But in the weightlessness of space for the first time, she was reduced once more to the clumsiness and incompetence of a baby. The co-ordination and precision built up over a lifetime spent in gravity were suddenly irrelevant, and to survive in her new environment she had to learn a whole new set of rules. Despite hours of training in a water tank – the nearest thing to weightlessness available on earth – it took her ten days to master the new conditions.

In space objects may not have weight, but they do have mass. Push something too hard and the momentum created will send it spinning out of control. Similarly, a mistimed nudge of the foot off the cabin wall will launch the astronaut off at the wrong tangent, and the target of the move will be missed. The instincts that apply in normal life must be overridden and replaced with new ones. 'To begin with,' says Marsha, 'it requires conscious thought. Then over time it becomes increasingly automatic. After a while we can move things in weightlessness without thinking about it – from recapturing a rogue toothbrush to shifting huge things out in space.'

Sure enough, on her second mission, she was able to adjust to weightlessness in only six hours. Like a baby learning to walk, her brain had absorbed the results of earlier trial-and-error attempts and stored the successful sequences.

Not many of us will make it into space, but we all make use of the brain's extraordinary learning abilities. Learning to drive a car, operate a machine or play a new sport are all reminders of that journey we once made from helpless baby to skilled, competent adult.

species. *Homo sapiens* is neither particularly strong, nor fast, nor nimble nor well adapted to heat or cold. Rather than being a specialist it is a generalist, eating from a wide selection of foods and adept at a range of activities – such as running, climbing, swimming, seeing, hearing. Humans are outclassed in every one of these by the experts (cheetahs, geckoes, seals, eagles and bats for those five attributes), and are out-run, out-swum and out-climbed by numerous others. But that all-round ability is a major human attribute, a key asset in the adaptation to so many different environments.

The brain revolution

It is the human brain that has got our species where it is today. Together with a clever hand and the ability to speak, both of which need a good brain, it was the dramatic evolution of humanity's cerebral hemispheres that permitted this single species to become the most accommodating, wide-ranging, intelligent, vociferous, creative and potentially destructive species the world has ever known.

Three million years ago the biggest brain among our ancestors measured 450 cubic centimetres (equivalent to modern chimpanzees and gorillas, and to the engine capacity of very small cars). Later *Homo habilis*, the so-called handy-man, had a 750 cc brain. *Homo erectus*, fire-user and also tool-maker, had a 900–1100 cc brain about 1.5 million years ago. Neanderthal man, who only recently became extinct, was even better endowed than modern humans, having, on average, a 1500 cc brain. Average modern brains are nearer 1400 cc for males and 1250 cc for females. In proportional terms the smaller female organs form 2.5% of the body while the bigger male brains make up 2%. Many animals have even better body/brain ratios: in some monkeys it is nearer 5%. And various animals have bigger brains than humans, such as elephants (four times as big) and sperm whales (six times larger).

But brain function is not dictated solely by size. Some intelligent humans have brains twice the weight of other intelligent humans. (The writers Anatole France and Ivan Turgenev are often cited, their respective brains being measured at 1000 cc and 2000 cc.) The human brain is still a fairly diminutive part of the body, despite its massive size increase during the course of evolution. Standard human beings have 13 times as much muscle, much more bone (forming 206 components), much more skin (1.9 square metres), more intestine (8.5 metres), over three times as much blood, and very slightly more liver. The heart, kidneys, spleen, pancreas, and lungs are the only major organs weighing less than the brain.

However crucial our brain enlargement, it was not the first of the various evolutionary changes which now identify our species. An upright stance arose

much earlier, when brain capacity was still modest and ape-sized. The astonishing discovery (in 1975) of hominid/human-like footprints at Laetoli, 50 kilometres south of the famous Olduvai Gorge in Tanzania, showed this to be true. The footprints were made 3.5 million years ago by upright individuals walking across some newly deposited volcanic ash. (At first it was thought the prints resulted from two pairs of feet, but had been imperfectly made. The confusion was immediately clarified when a third pair was postulated, walking behind in the imprint of the others – as apes often do.)

Everything that followed from bipedalism – a narrower pelvis, larger legs, different feet, a flatter thorax, a more upright head – was therefore in place, to some extent, *before* the brain started upon its remarkable growth. The upright stance may have made a bigger brain more necessary, with a different life-style presenting different problems, but it certainly freed the forward limbs from the dual tasks of bodily support and locomotion. They were abruptly available for a whole new range of occupations, permitting the age-old pentadactyl (five-digit) limb to develop into modern hands. An upright stance also caused the female pelvis to become narrower; easier for walking, harder for childbirth. It also meant humans were equipped with a spinal column which had originally evolved to exist horizontally, hence our problems with backache in general and slipped inter-vertebral discs in particular.

Size (and weight) of brain tissue is less important than the degree of convolution. Such folding, causing ridges and grooves, leads to a greater surface area, and this is highly relevant to the brain's ability. So too is neural organisation, the inter-connections between the millions of nerve cells within the 1.4 kilograms of human brain. These cannot be mapped in a living brain, and there is no way – yet – of discovering even in a dead brain where its former owner would have rated on the IQ scale. (Albert Einstein's brain has been sitting in a jar in Wichita, Kansas, since his death in 1955. Perhaps it will be examined one day

● *Left* An adult chimpanzee's brain, which grows no more after the chimp's first birthday, is about one third of the size of an adult human's brain. Men have, on average, larger brains than women, but women's brains occupy a greater proportion of their bodies.

to discover if his brilliance is detectable.) Neither is there any way – for the uninitiated – to tell that the brain is the organ of intelligence. Aristotle, the first to ascribe functions to organs, considered the brain to be a cooling mechanism, with a runny nose being the release of cooling fluid. He concluded that the heart was the seat of thought because it beat faster from excitement or fear.

The brain is divided into distinct structures, all with different roles. At the base of the brain, above the spinal cord, is the brain stem. This area controls involuntary movement, posture, and all the sub-conscious tasks required for the general maintenance of life. Above the brain stem is the cerebrum, or fore-brain, which includes the highly enlarged cerebral hemispheres.

These hemispheres are, in essence, where *Homo sapiens* resides. Most of the brain's appetite for oxygen is generated here: they are 16 times more demanding, by weight, than any other tissue. They are separated from each other by the longitudinal fissure, linked only by a thick bundle of nerve fibres running from one side to the other. The cerebrum, in particular the cerebral hemispheres, is the part of the brain that most distinguishes humans from animals. As we have already seen, it was this portion that enlarged so dramatically from early to modern humans. To have enlarged the hind-brain would have been too disruptive, partly because its components abut on to each other and have important tasks to perform. Far better to push forward the brain's front portion – to build an extension rather than modify the house.

Despite their homogeneous appearance, each cerebral hemisphere is divided into four lobes. The functions of the lobes are by no means clearly mapped out. However, we do know that they are the seat of our 'higher' activities – voluntary movement, coordination, the recognition and interpretation of sensory inputs, and abstract thought and emotions. Different functions are located in specific areas or centres, but most involve complex interactions between several centres. New imaging techniques are helping to unravel, bit by bit, more of the mysteries of cerebral function.

It is also bewildering to contemplate the many billions of neurones (nerve cells) within the unexciting form of a human brain. Each cell has many hundreds (or even thousands) of extensions, which can each link via gap junctions called synapses to any one of the similar extensions from neighbouring neurones. The eventual number of connection possibilities is not infinite, but it might as well be. The actual figure (whatever it is) is colossal, and well beyond the imagination of most of us. However, it is this bewildering network which is responsible for our intelligence. The number of neurones is a factor, but the number of connecting pathways is more important.

The brain's evolutionary growth during the past three million years has been called 'explosive'. Gaining less than one kilogram in that length of time might appear sluggish, but it means about 8 cubic millimetres more brain tissue per generation, or to put it another way 150,000 extra brain cells in every 25 years. The additional number of possible pathways and inter-connections developed in every generation as our species evolved is therefore phenomenal. A question inevitably arises: why was there such an increase in brain power, bearing in mind that brains are costly in energy consumption, requiring far more oxygen – 20% of the available total – than any other organ? The modern brain receives about 30% of the blood leaving the heart, or roughly its own weight in blood every minute. What advantage did this increasingly powerful, and demanding, brain provide for our ape-like ancestors?

Until quite recently these people-ancestors were hunter-gatherers, much as chimpanzees are today (and there is no question – far from it – of chimpanzees taking over the world). A huge brain is not essential to learn and remember where fruits are ripening, or nuts are maturing, and many animals are excellent hunters with only modest brains. Yet the human brain went on developing until it could create and remember symphonies, invent writing, study science, weigh the planet on which it was born, and even devise a means for leaving it. This full potential was acquired long before – some say 100,000 years before – any such activities were performed, while humans were still hunting and gathering, much like various traditional societies that exist today. But why? As the biologist Alfred Russel Wallace phrased it more than a century ago: 'An instrument has been developed in advance of

THE BLOOD AND CIRCULATION

The blood is a fluid organ – not some inert liquid but a living tissue made up of cells like any other. It is first and foremost a transport system, supplying the body tissues with oxygen and nutrients, carrying away their waste, and bearing the hormones and other messengers that regulate their functioning. It is also a coolant, taking away metabolic heat and dispersing it from the skin and lungs. Finally it is the organ of immunity, its white cells being responsible for fighting off foreign invaders.

Blood consists of red cells, white cells and cellular fragments called platelets, suspended in a watery, protein-containing base known as plasma. Red cells are the most numerous and their function is to transport oxygen. This is carried on haemoglobin, a large protein molecule containing iron. Haemoglobin picks up oxygen atoms at the lungs, where oxygen concentration is high, and drops them off again in the tissues, where oxygen concentration is low. When oxygenated, haemoglobin is bright red, but once the oxygen is off-loaded it turns blue. Blue blood is visible in the veins beneath the skin, but we never see blood directly in this unfamiliar state; as soon as a vein is cut the haemoglobin encounters oxygen and turns red again.

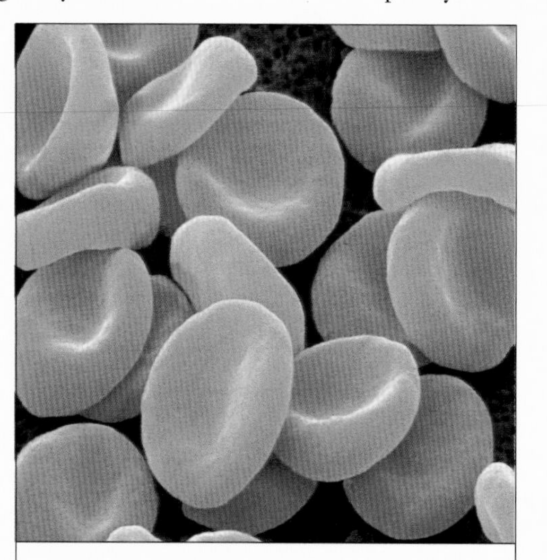

● *Above* Red cells (erythrocytes), whose function is to transport oxygen from the lungs to the tissues, are the most abundant type of cell in the blood. There are about 5 million in each cubic millimetre, accounting for 40%–45% of the blood volume.

● *Right* A cross-section through an arteriole (a very small artery). Its muscular wall contracts and expands, thus regulating arterial blood pressure. Red blood cells can be seen travelling through it.

Transporting the blood around the body is a branching network of vessels which, if placed end to end, would stretch for almost 100,000 kilometres. These are divided into: arteries, which carry oxygenated blood away from the heart; capillaries, which exchange the blood's cargo with the tissues; and veins, which take deoxygenated blood back to the heart. The largest vessel is the aorta, the artery emerging directly from the heart, whose diameter is 2–3 cm. The aorta branches into a series of major arteries, which in turn branch into small arteries (just visible to the naked eye). These diverge again into microscopic arterioles, and finally into a bed of capillaries just five one-thousandths of a millimetre wide – so narrow that red cells must pass through in single file.

Instead of having a muscular coat like arteries and veins, capillaries have very thin walls with loose joins between the cells that allow fluids and small molecules to leak out. Oxygen and nutrients diffuse through and are taken up by the surrounding cells, whose wastes diffuse in to the capillary. As the blood flows onwards the network of vessels gradually reunites, first into venules, then veins, and finally into the great vena cava which re-enters the heart. To prevent blood from flowing backwards, the veins are fitted at regular intervals with small, one-way valves. If these become leaky the blood forms pools, which are visible under the skin as varicose veins.

Driving the blood around the system is the four-chambered heart. At rest it pumps about 5 litres a minute, equivalent to the entire volume of blood in the body. In strenuous exercise its output can reach 20–30 litres a minute. Keeping the heart beating takes up 5–10% of the body's total energy budget.

The heart is made of a special type of muscle that has an inherent tendency to contract. The contractions are stimulated and coordinated by signals from the pacemaker, a group of electrically active cells. Impulses spread through the heart causing its chambers to contract in a specific sequence. The heart also receives inputs from the nervous system, which is why it speeds up when we are frightened, excitable or emotional.

Humans and all mammals have what is known as a double circulation, meaning that blood passes through the heart twice. On the first pass it is sent to the lungs to collect oxygen and get rid of carbon dioxide. From there it returns to the heart and is pumped at high pressure out into the arterial system. The arteries are elastic in order to cope with the surges of blood. As we age they become stiffer, causing the blood pressure to rise and making the heart work harder.

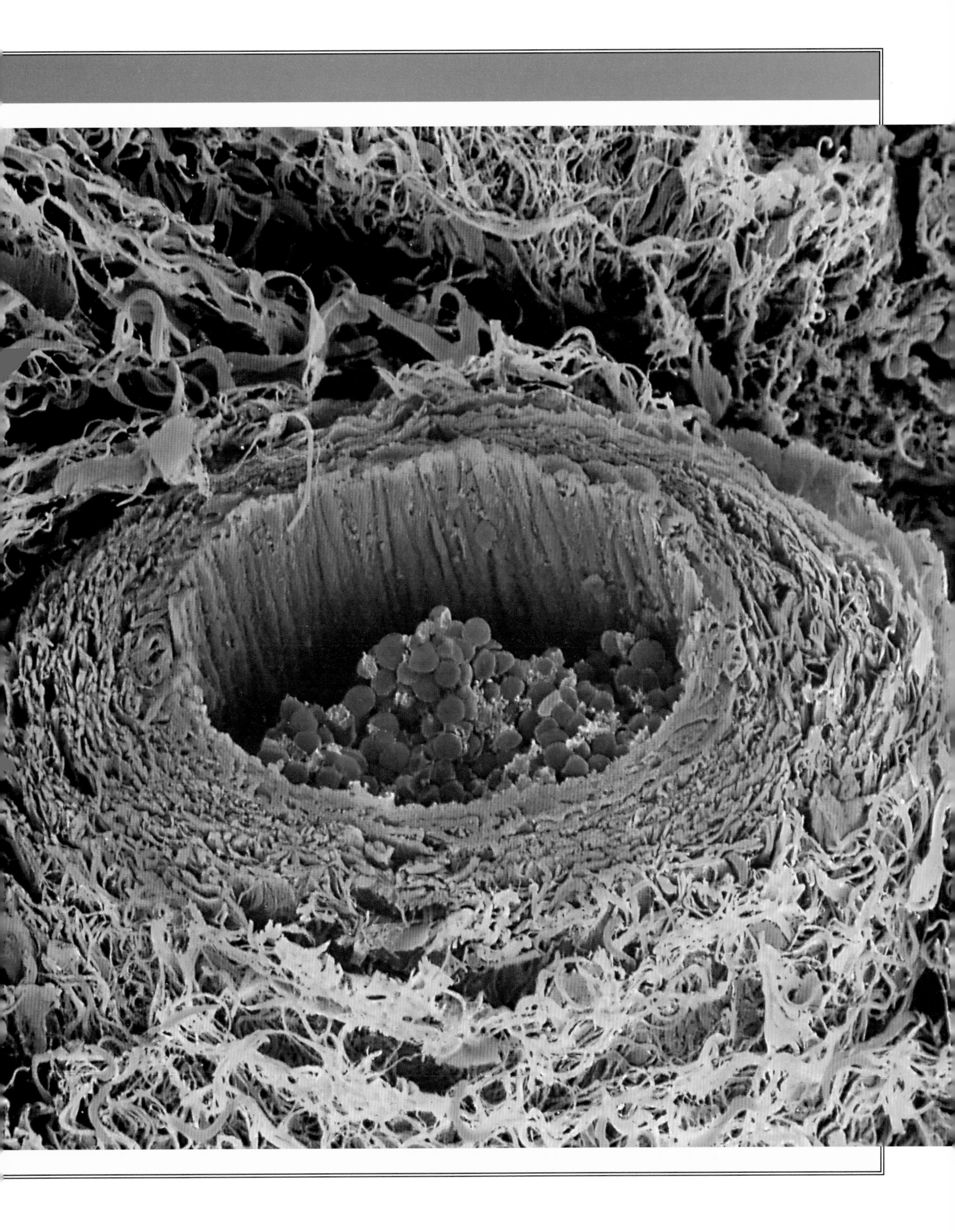

the need of its possessor.' Or, as Arthur Koestler wrote more recently – and more bluntly: 'It is the only example of evolution providing a species with an organ which it does not know how to use; a luxury organ, which it will take its owner thousands of years to learn to put to its proper use – if it ever does.'

One explanation is that, if increased stature is desirable (and is being achieved via evolution), all bodily components – whether liver, pancreas, kidney or brain – will also increase in size. Another is that cleverer humans need cleverer brains to deal with each other. An increasingly complex society makes increasingly complex demands, and if warfare is involved (as practised even by chimpanzees) an increasingly cunning human enemy has to be overcome. In short, the most complicated thing that ancient humans would meet in their lives was not tools, or prey, or predators; it was other people. Even bats support this argument. The solitary kinds have small brains; social bats, such as vampires, have larger ones. They need to thrive among all the interactions of a complex society.

The tool of language

No other societies are quite so entangled as human ones. We spend 90% of our talking time discussing other people, commenting upon their relationships, scheming, outwitting, asserting. This has probably been the case ever since talk began, or even before that time, with gestures, grimaces, cringes and swaggers pre-empting words. Chimpanzee groups, for example, spend tremendous time and energy on hierarchy in all its forms, on who is permitted to do what with whom, and how and when and where.

Homo sapiens is the talking ape, but no one knows when speech began. Signs, gestures and sounds are able to convey information, instructions or attitudes quite effectively. There are dictionaries defining tens of thousands of human winks, nods, glances, shrugs, nudges, becks, grips, snarls, smirks, lip-curlings, finger-waggings and seemingly endless communicating possibilities without a word being spoken. Sir Richard Paget, 1869–1955, decided that humans could make 700,000 gestures, a figure to compare with the 4000 words most of us use in ordinary talk.

The more that behaviourists have learned about animals the greater is their amazement at the quantity of information imparted without the gift of speech (for example, vervet monkeys have three distinct alarm calls for three different kinds of snake). But the ability to talk – and the corresponding ability to disentangle the speech of another – is a tremendous advance.

Speech could not occur until there was both a brain to handle it and a larynx to express it. A newborn human baby cannot speak, partly because its larynx is not yet properly located (being in the same position as in all other mammals). This enables it to suckle and breathe at the same time. During those earliest months every human baby is a nose-breather, unable to use its mouth for inspiration. If the baby's nose is blocked it will die for lack of air. Both its larynx and its jaw have to move downwards before mouth-breathing becomes possible. The mammalian larynx did not initially evolve for speech, but as a valve at the head of the air tract to the lungs, a feature developed by early land-based amphibians in response to the need for breathing air. Only in frogs, toads, some lizards and most mammals is it now a vocal organ. Birds have a larynx, but this is without vocal cords. Their song comes from a syrinx, similar to a larynx but much lower down the air passage and therefore nearer the bronchi and the lungs.

The vocal cords are wrongly named, being folds of tissue rather than anything resembling string. These folds create a wedge-shaped opening in the larynx. All air to and from the lungs goes through this hole – 17 mm long in men, 12 mm in women – and the 'cords' only come into play during vocalisation. The source of sound is not the actual vibration of the cord tissue, but the repeated interruption of air-flow caused by its valving action – another reason for not calling them cords.

At six to eight months human vocalisation can begin. Every baby makes noises before this – quite apart from crying. There is babbling (after about eight weeks) and cooing, with all babies initially babbling in a similar fashion whatever language is being used nearby. Such babbling may stop at nine months, as if the baby has suddenly realised the purpose of surrounding adult talk. At that time the young human starts to imitate adult sounds, and may even use proper words. The first words are spoken anywhere between eight and 30 months.

At some point every infant must make the extraordinary mental leap of realising that there is sense to the adult chatter that it hears. It then embarks on a spectacular learning curve. Only about a quarter of an 18–month-old's talk makes sense, improving to two-thirds at two years and 90% at three years (although there is considerable variation from child to child). During this time the child's brain has a tremendous and never-to-be-repeated aptitude for learning sound recognition and grammar – as explored in chapter three, 'Childhood'.

Language could not have developed without a larynx (or some such device) but the brain must also have evolved sufficiently to organise the sounds. So when? The earliest man-made tools (to have survived) were fashioned about three million years ago. They were crude, being little more than pebbles with a flake or two knocked off to provide a cutting edge. For hundreds of thousands of years afterwards there was little advance, and the hand-axes which then followed did not change for another great span of time. It is therefore tempting to believe that speech, with all its possibilities for handing on acquired information, could not have co-existed with such modest progress in tool-making. Then, much more recently and quite suddenly, exquisite stone implements began to be made. Within a short span of time humankind had started upon the astonishing advance which has led to our current situation.

Was it the arrival of speech that both initiated and then paralleled this explosion? It does seem plausible, even if good evidence is lacking. There is also the conundrum that the human brain has not increased in size during the past 100,000 years, just when its superior ability has been exercising its potential – by domesticating animals, encountering (and surviving) an ice age, treating food, making shelters, beginning agriculture, selecting suitable varieties of useful plants, creating art, and initiating what has been called the neolithic revolution. That revolution led on to city dwelling, to writing, to government and trade, – to the kinds of societies that are our lot today, and that are still evolving. It could not have happened without the human brain, the 50 billion (or so) neurones with which each of us is now endowed.

The hand and handedness

Neither could the neolithic revolution have happened without the human hand. The upright stance liberated our ancestors' fore-limbs for further development. Chimpanzees and gorillas walk on their knuckles, and orang-utans use their fore-limbs primarily for holding branches, but the human hand became available for transformation into tool-maker, manipulator, and dextrous 'extension' of the brain. As it changed the brain imposed a strange system of preference that shaped not only our manual abilities but our very culture.

Thomas Carlyle, forced to write with his left hand when his right became unusable, called it the oldest human institution of them all. There is some evidence that Australopithecines (very early hominids) used the right hand more than the left – in attacking skulls. Cave painters (relatively very recent) favoured the right, and neolithic skeletons are more frequently damaged on the left, suggesting right-handed opponents. The Bible has 1600 references to rightness and righteousness, almost all hostile to the left side, the cack-handed, devilish, gauche and non-dextrous side. 'Let not thy right hand know what thy left hand doeth,' wrote St Matthew, along with 'And he shall set the sheep on his right hand, but the goats on his left.' The words 'sinister' and 'sinistral' (meaning left-handed) have the same Latin root.

Spiral staircases in castles favoured right-handed defenders. Middle eastern and Asian eating is right-handed, leaving the left free for toiletry. Four-fifths of all madonnas in Renaissance paintings are cradling their infants left-handedly,

THE CLEVER HAND

Next to its brain, the hand is probably *Homo sapiens'* greatest asset. Other species may run faster or have keener senses, but none comes even close to matching our manual skills. The human hand is capable of both brute strength and incredible finesse. It carries out hard labour yet is sensitive to nuances of texture and form, so much so that it can serve as a substitute eye to the blind.

Much of its prowess is attributable to the brain that controls it, coordinating its actions with signals from the eyes or feeding in complex instructions at an almost automatic level. Watch any pianist or typist to admire the brain's orchestration of hand movement.

The ability to use the fingers independently rather than as a clumsy bunch is a key part of human dexterity. In addition to this, there are two forms of manual grip which outclass the capabilities of any other creature.

The first makes use of the opposable thumb, a feature shared – although less markedly – by various other primates. The human thumb can touch, with varying degrees of force, every portion of every other digit on its own hand. The second, and possibly more important, skill is the double grip. One hand can hold two snooker balls separately, for example. With one ball gripped between the hand's last two fingers and its palm, the second ball can be held, most delicately if need be, between forefinger and thumb. Having strength and delicacy intermingled in this way has permitted the human hand to be more gifted, more capable and more manipulative than any limb ever seen.

Holding two balls independently, or unscrewing a pen-top with one hand, may seem trivial abilities. But these two forms of grip put humans in the top class for tool-making, for weapon-making, for doing whatever their clever brains ask them to. A chimp could never make a hand-axe. Extremely sensitive finger-tips, and the whorls, loops and ridges of finger-prints (unique to every individual, and helpful for gripping) must also have contributed to human manual skill. Unfortunately, evidence on the hand's evolution is lacking, as fossils of the small bones of hands are not so easily found as the more rugged remnants of a skull.

Even if fossil hands were abundant, the mere bones would not necessarily reveal much about dexterity. Fortunately there are hand-made artefacts in lieu of fossil hands. The pebble tools were fashioned for a million years. Hand-axes then took over, initially (and largely) in Africa. Once again, with thousands of generations making more or less the same objects, the human hand and its guiding brain were presumably still primitive. Perhaps the important phenomenon of handedness, concentrating dexterity on one side, had not arrived.

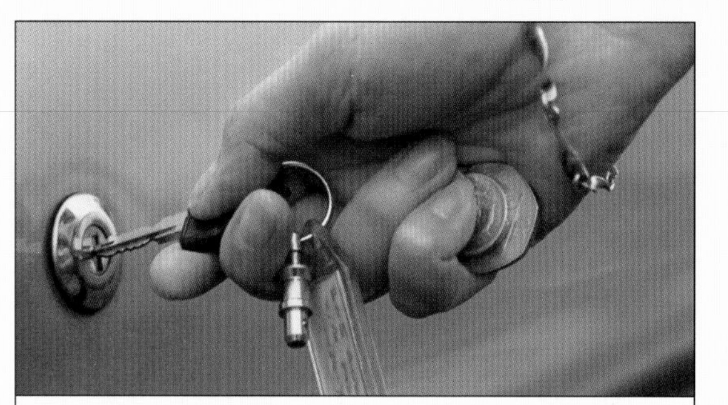

● *Above* The human hand is unique in having two distinct grips. In the above example, not only can several items such as coins be held firmly in the palm with several fingers but also, simultaneously, the opposable thumb can be used to hold and turn a car key.

● *Below* Progressive bone growth in young hands. At age 2.5 (*left*) there is proper calcification only in the middle of the longer bones. At 6.5 (*centre*) calcification has progressed, and the spaces between the bones at the joints have narrowed. At age 19 (*right*) the bones have fully calcified and the joints have virtually closed.

permitting right hands to be available for other tasks. Even the human scalp whorl is more frequently on the right and usually gyrates clockwise. About 90% of modern individuals favour their right hand over their left.

So why this preference? Some palaeontologists argue that handiness, as with early humans like *Homo habilis*, could not have developed so successfully without handedness. Dexterity is much more refined on one side – try writing, screw-driving, hammering, sawing with the wrong hand. The brain's left half (which controls the body's right and, for most people, favoured side) is also the half for speech control. These two human abilities are therefore closely located in the brain, which may or may not be significant. All four of the big apes tend to be handed, but these closest of relations are equally divided between right and left. Many other primates are undecided, using both hands without a preference. A few humans are truly ambidextrous, equally competent (rather than incompetent) on both sides. Edwin Landseer, the artist, could draw one animal with his right hand while drawing another and different animal simultaneously with his left.

An oddity about handedness is the high frequency of deviation from the norm (if the word 'deviant' is acceptable to left-handed individuals). Most human aberrations – such as a right-sided heart (dextrocardia), three breasts (polythelia), or six fingers (polydactyly) – are rare, but left-handedness is both common and without apparent disadvantage. It occurs all over the world, in all races, and is almost equally divided between the sexes (present in 12% of males and 10% of females, according to a recent survey).

There is a genetic influence on handedness. If one parent is left-handed there is twice the normal likelihood that the offspring will be. If both parents are left-handed this likelihood is three times the usual rate. With one-egg (identical) twins about 80% are also identical in their handedness, be it right or left. More one-egg twins are left-handed (15%) than are two-egg twins (11%).

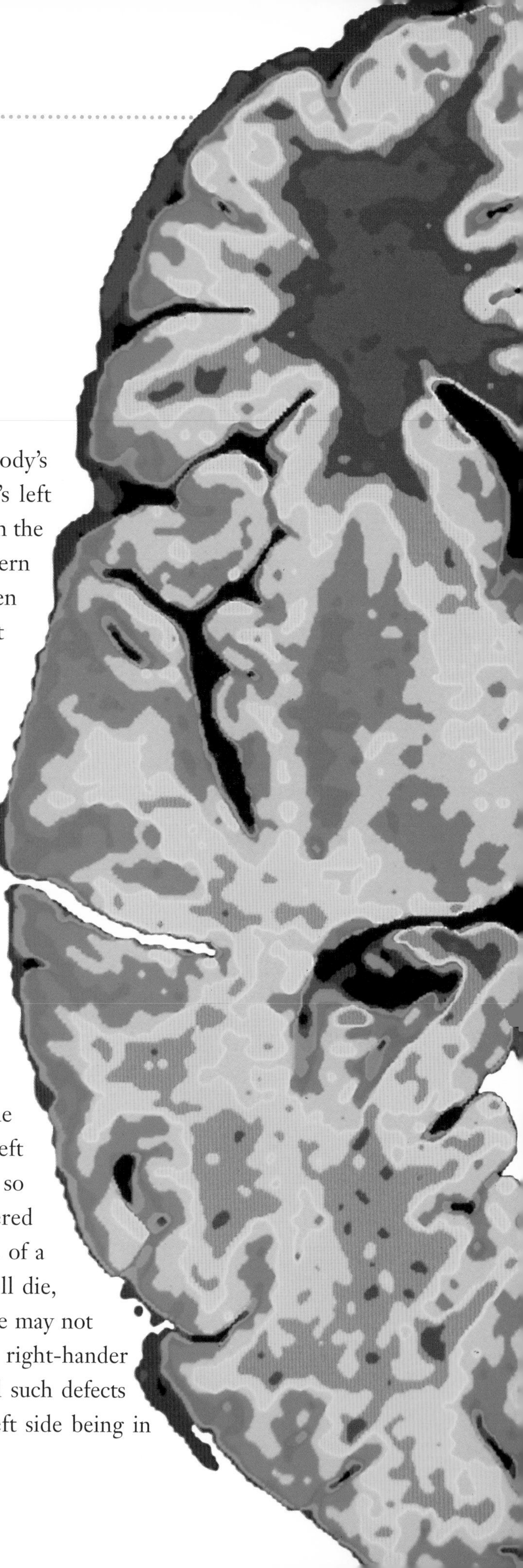

The split brain

The fact that both speech and the body's right side are controlled by the brain's left half is one indication of the way in which the brain enlarged from primitive to modern humans. In essence the growth has been achieved by a pushing forward of the left and right cerebral hemispheres. They have each swollen, but independently of each other. The brain's two halves became increasingly separate the more each hemisphere bulged forward (a bit like two small boxing-gloves, joined only at the wrist). There are bundles of nerves connecting the two halves, but they are not substantial. There is far more communication within each hemisphere than there is between them.

In all the other double organs – kidneys, breasts, testes, ovaries, lungs – the two sides perform precisely the same function. The brain is unique in that left side and right side function differently, so much so that the left side is considered dominant even for left-handers. Victims of a stroke destroying the dominant side will die, whereas similar damage to the other side may not be fatal. Damage to the right side in a right-hander rarely leads to speech defect. Almost all such defects are caused by left-sided damage, the left side being in control of language. The asymmetry is even present at birth. In general babies turn to the right four times as often as to the left. Their grip, when a little older, is more tenacious with the right hand. Balancing something with a right finger is easier than with the left, but not so if the balancer is speaking simultaneously. Amputations above the right knee create more phantom limb pain than those above the left knee. The left breast is more sensitive than the right. The right-prejudice even extends to drawing faces. We tend to favour one profile over the other. In short, nothing seems to be equally divided between the two halves.

With both speech and handedness usually controlled by the brain's left and dominant half, the right side is available for different tasks, as well as controlling the body's left side. It is more important for spatial activity, for the expression and reception of emotion, for aspects of music, and for facial recognition. The left side is the verbal side, and the right non-verbal. This can be tested with a cooperative friend. Stand straight in front of this person and ask a verbal or arithmetical question. The

Left A magnetic resonance image (MRI) of a normal brain in horizontal section. The split between the left and right hemispheres is clearly visible.

friend's eyes will probably be averted, and look to the right. A spatial question – for example, what is directly beneath the upstairs toilet? – should have the eyes moving to the left. The eyes move to reduce the visual input to that half of the brain attending to the question – to the right for verbal questions, and to the left for spatial ones.

The asymmetry can also be investigated in the laboratory. Brain waves, such as the dominant alpha rhythms, are detectable by electroencephalograms (EEGs, which record electrical activity using electrodes placed on the head). The half of the brain being exercised registers a greater change in alpha activity. A verbal query will elicit more alpha response in the brain's left half, and vice versa with spatial queries. One further, and exceptionally neat, experiment with an EEG is even more dramatic. To whistle a tune leads to more alpha activity within the right hemisphere. To speak that tune's words increases alpha rhythms on the left side. But to sing the song, using music *and* words, influences alpha wave activity on both sides. (Needless to say, with so much asymmetry in the brain, even the underlying alpha rhythms are unevenly created, being slightly more pronounced in the right hemisphere.)

Brain waves were first discovered in animals in 1875, but the subject was not properly explored until 1929. The psychiatrist Hans Berger then discovered four principal rhythms in humans: alpha, dominant when the mind is resting (8–13 cycles a second); beta, most noticeable when the brain is attentive (more than 13 cycles a second); theta, related to drowsiness (4–8 cycles a second); and delta, usually present only in deep sleep (less than 4 cycles a second).

Brain signals and connections

These waves are all recording natural happenings, but have not been as informative as was originally hoped. After all, they are the 'sounds' of a brain at work. They can certainly show if their originators are adult, child, asleep, awake, dreaming, open-eyed, shut-eyed, suffering an epileptic attack, dead, or even have dead areas within the brain (perhaps caused by tumours). But they do not illustrate how the brain works, or what precisely is generating them. They are merely the electrical background caused by 50 billion neurones carrying out their business.

Each nerve cell can fire a very small electrical pulse, and does so according to inputs received from other nerve cells. Individual pulses are weak – the entire brain only creates about 20 watts of power – but each firing has to be sufficient to energise the next cell, or group of cells. Signal transmission from one cell to another is achieved chemically rather than electrically, with a chemical known as a neurotransmitter released at each nerve ending. The transmitter crosses the synapse (gap) to the next neurone and becomes lodged in a specific receptor site. This then activates a chain of events in the second neurone which causes it to fire and thus pass the impulse on. Within the human brain there are scores of such transmitter chemicals, some exciting their target neurones and some inhibiting them. Many psycho-active drugs exert their action by interfering in some way with the neurotransmitters. The nerve impulse itself is generated by changes in the nerve cell's outer membrane, which influence the flow of electrically charged ions into and out of the cell.

The brain
- more than the sum of its parts

How does the passing of an impulse enable us to remember faces, think consciously, behave with intelligence? 'How indeed!' is one response, for these astonishing processes are still poorly understood. The human brain is having difficulty in unravelling the fundamental processes of its own working. We have invented computers, far speedier than the brain at many tasks, but their procedures are not analogous with neurological activity and therefore do not help us to understand the workings of the brain.

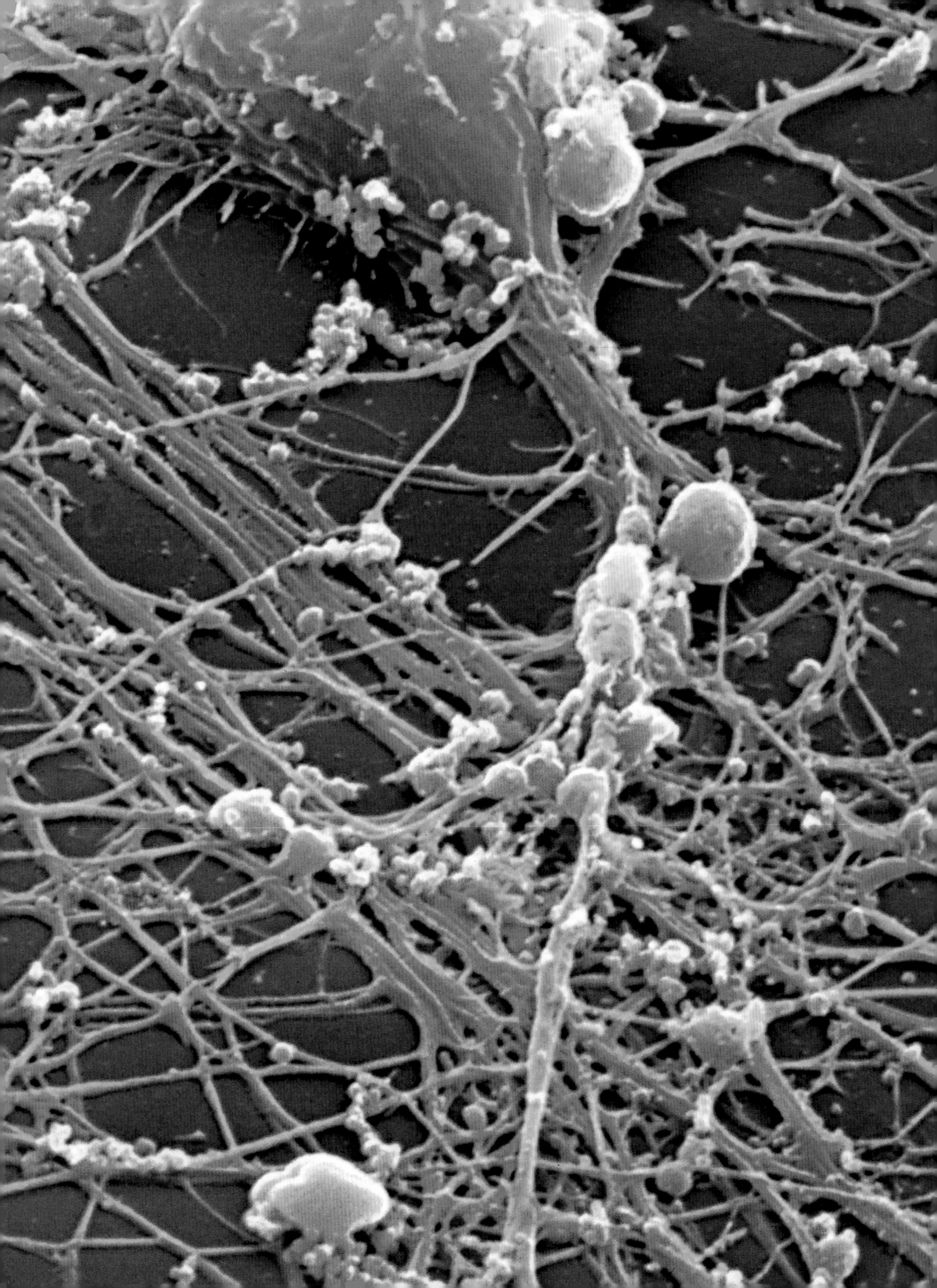

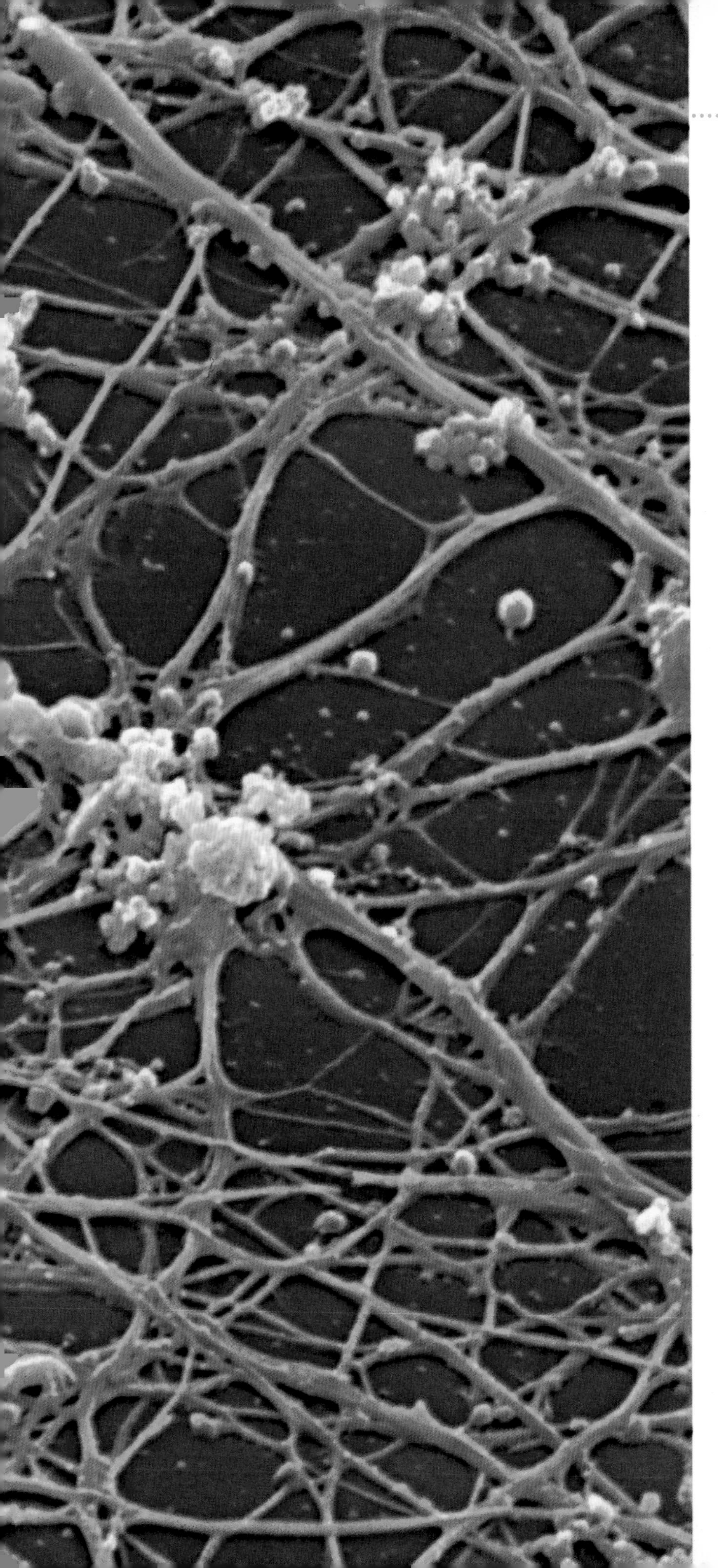

Left Branching nerve fibres form a dense network in the brain. The large object at the top left is a cell body with fibres leading out from it. All around are outgrowths from other brain cells.

A better comparison is with the activities of termites. These creatures show a formidable degree of organisation, from communal foraging, and even waging war, to the construction of enormous mounds complete with columns and ventilation shafts. Despite every indication that an intelligence is at work, there is no master-mind behind the colony's activities. The queen is not the architect – her brain is actually smaller than those of her workers. The individual workers, although industrious, are capable of little. It is only from the colony as a collective that the extraordinary feats of organisation emerge. The colony is more than the sum of its parts, its complexity emerging from the combination of individuals in a way that could not be foreseen by observing an individual insect. So it is with the brain, whose capacity for reasoning, memory, emotion and the rest of its activities could not be predicted from the simple properties of the individual neurone.

It is amazing that simple animals with far smaller numbers of neurones than the 50 billion we possess are still capable of quite complex existences. The parasitic worm *Ascaris* always has 162 brain cells. With these it can learn, act on information received, and store facts within its memory. *Apis mellifera*, the honey bee, is considerably more capable than the intestinal worm, but still only possesses some 7000 nerve cells. We are often astonished at our brilliance, but that bee can build, clean, repair and ventilate a complicated hive, inform other bees of the *direct* line to a food source (even if an indirect line was taken to find it), steer a course pioneered by others, recognise friend or foe, know the time of day (to within 30 minutes), estimate the sun's angle (to within 3 degrees) – and collect nectar and pollen.

The suspicion therefore arises that humans should be even more capable than they are, owning such a massive brain by comparison. The bee's brain occupies 0.74 of a cubic millimetre as against the human capacity of 1400 cubic centimetres. Do our brains have abilities that our species has not yet discovered, or has forgotten how to use?

Learning not to think

Humankind has been happy to describe itself as the thinking ape, but much of our individual progress is achieved by *not* thinking about what we are doing. 'Learning,' wrote Peter Medawar, biologist, 'is learning not to think about operations that once needed to be thought about.' Similarly Alfred Whitehead, philosopher, wrote: 'It is a profoundly erroneous truism ... that we should cultivate the habit of thinking what we are doing. The precise opposite is the case.' Think about walking downstairs and, instead of doing it more efficiently, it is easy to stumble as a consequence. Think about any 'automatic' act and confusion may result.

The part of the brain devoted to these automatic movements is called the cerebellum, meaning 'little brain'. This takes over from the conscious part of the brain and stores standard movement routines. Once these movements are learnt, they can be instigated with a voluntary thought and then proceed like clockwork. The cerebellum means we can sit, stand, walk and run without having to make conscious calculations about posture and balance. Compare this with a toddler taking its first steps. The automatic movements and adjustments have not yet been committed to the cerebellum, so every pace requires a huge feat of concentration and conscious coordination.

Most mammals have cerebellums about as sophisticated as our own, at least in some functions. In the rat the brain is virtually all cerebellum, and its range of automatic movements makes the animal superbly adapted to its environment. Unlike many other parts of the brain, the cerebellum has changed very little during evolution. However, humans have a unique ability to extend this sense of automation towards tools and pieces of machinery, enabling levers, wheels, pedals and buttons to serve as sophisticated additions to their limbs.

THE IMMUNE SYSTEM

Every hour of every day, our bodies are under attack from alien life-forms. Some just use our skin, mouth or gut as a convenient home, and cause us little trouble. Others are bent on hijacking the whole system for their reproductive needs, even subverting our DNA to their own ends. The invaders are bacteria, viruses and fungi, and without an effective immune system they would soon overwhelm us.

The body's first line of defence is the skin, which forms an effective barrier to the entry of microbes. Once they have gained access, whether through skin damage, inhalation or ingestion with the food, the task of defence falls to the white blood cells. These come in many different types, each with a specific role. Collectively, they are known as leucocytes. They are helped by a complex array of protein messengers, whose actions scientists are only just beginning to understand. Perhaps the best-known of these are the interferons, which are attracting interest as possible treatments for multiple sclerosis and cancer.

The key to our immune defences is the body's ability to differentiate self from non-self, i.e. to recognise 'foreign' matter. It does so by responding to chemical combinations. These may be as simple as the nickel present in a cheap pair of earrings, but in the case of organic matter the response is mainly to proteins carried on the outsides of foreign cells. These proteins, known as antigens, act as identification badges that are recognised by particular types of white cell.

White blood cells are involved in two types of defence, termed specific and non-specific immunity. Non-specific immunity is mounted when the skin or other tissue is damaged. Chemicals released by the damaged cells cause increased blood flow to the area and attract white cells called macrophages, which engulf and digest foreign organisms or particles. The swelling and pain we experience after injury is a sign that the immune response is in action.

Specific immunity is a learned response to a particular antigen. In 430 BC, a writer describing a great plague in Athens observed that 'the same man was never attacked twice – never, at least, fatally.' By the early seventeenth century, people were being immunised against smallpox by having pus from a victim's pox inserted into cuts in their skin. Today we routinely vaccinate against diseases using a dead or weakened preparation of the relevant virus or bacterium.

When a new type of antigen enters the body for the first time, it is recognised as an invader by the white cells. Over a period of about 20 days, other white cells produce large protein molecules containing chemical regions that recognise and are attracted to that specific antigen. These are known as antibodies. When they come across the antigen they stick to it, slowing down its movement and marking it out for destruction by macrophages. Because antibodies take time to develop, the new infection is able to take hold and the 'host' individual goes down with the disease. After recovery, low levels of antibody remain in the blood. If the particular antigen enters the body again, more antibodies are quickly produced to overcome it. The ability to make the antibodies is stored in 'memory cells', believed to reside in the spleen, liver and bone marrow.

The immune system is extremely complex, and is still far from being fully understood. The account given here is highly simplified: in reality immunity involves the interaction of several different types of white cell and many chemical messengers. The devastating consequences of AIDS, where the body can be overcome by microbes that are normally harmless, are caused by the loss of just one class of white cells.

● *Left* Macrophages (or scavenger cells) are a type of white blood cell found mainly in connective tissue. Their function is to engulf and destroy unwanted material such as invading bacteria and worn-out cells. The picture shows two macrophages and reveals their delicate external structure. The protrusions help them to fasten onto their prey.

● *Right* A macrophage (yellow) consuming an old red blood cell.

Occasionally the system becomes confused and starts attacking 'self' antigens. This is called auto-immune disease. For example, in juvenile-onset diabetes the immune system is thought to attack and destroy the cells in the pancreas that make insulin. Nobody knows why this should happen. In allergy a person becomes hypersensitive to a particular antigen, often one found on pollen grains, dust mites, or animal fur, or present in a food. The 'allergen' is recognised by a particular class of antibody that causes an immediate inflammatory response at the site of contact, resulting in symptoms like itching, watering eyes, swelling and asthma. In very severe cases the inflammatory messengers spread throughout the body, causing widespread dilation of blood vessels and a consequent drop in blood pressure that can be fatal. This rare condition is known as anaphylactic shock. Susceptible individuals can carry a syringe of adrenalin that will reverse the process when injected.

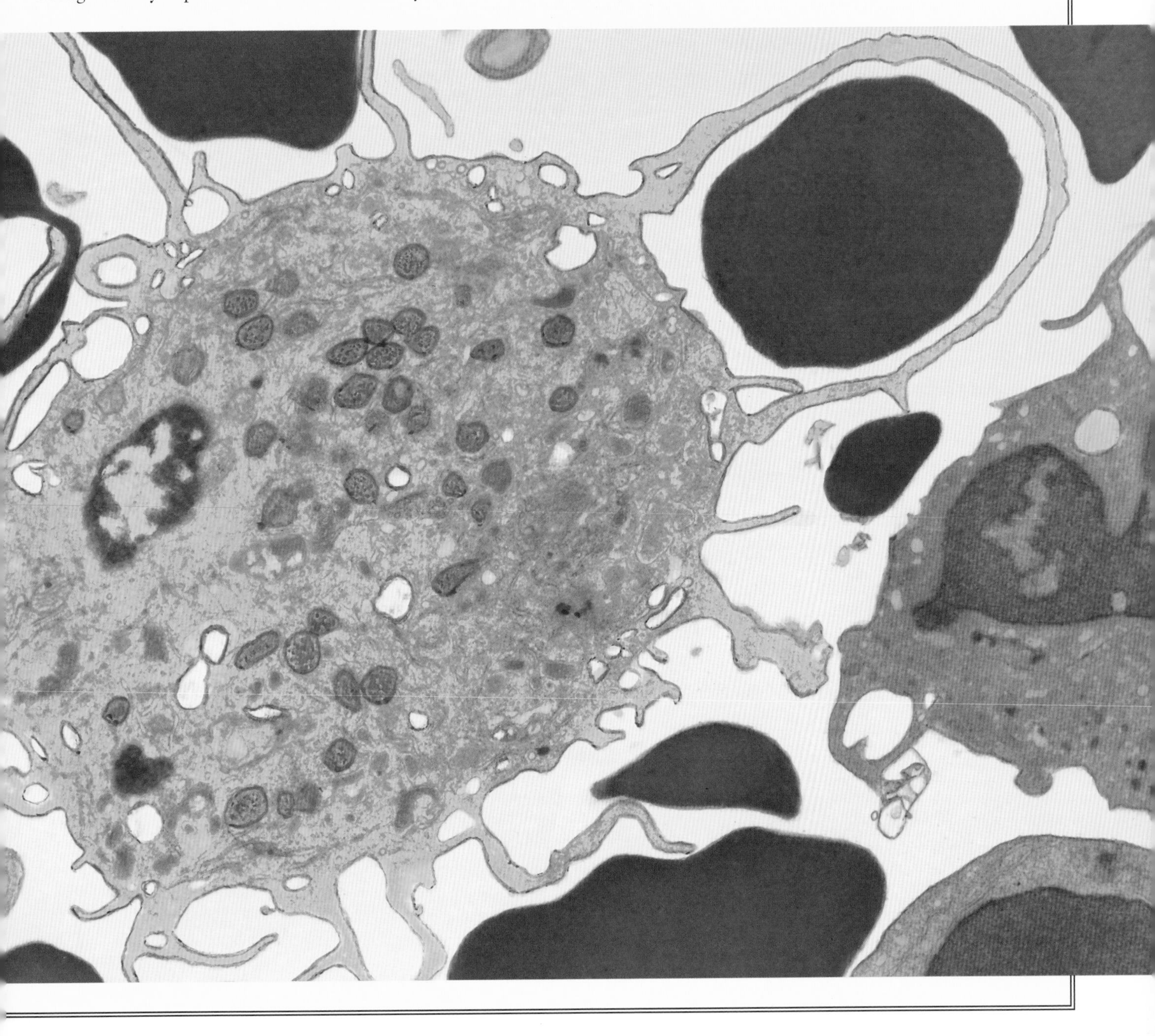

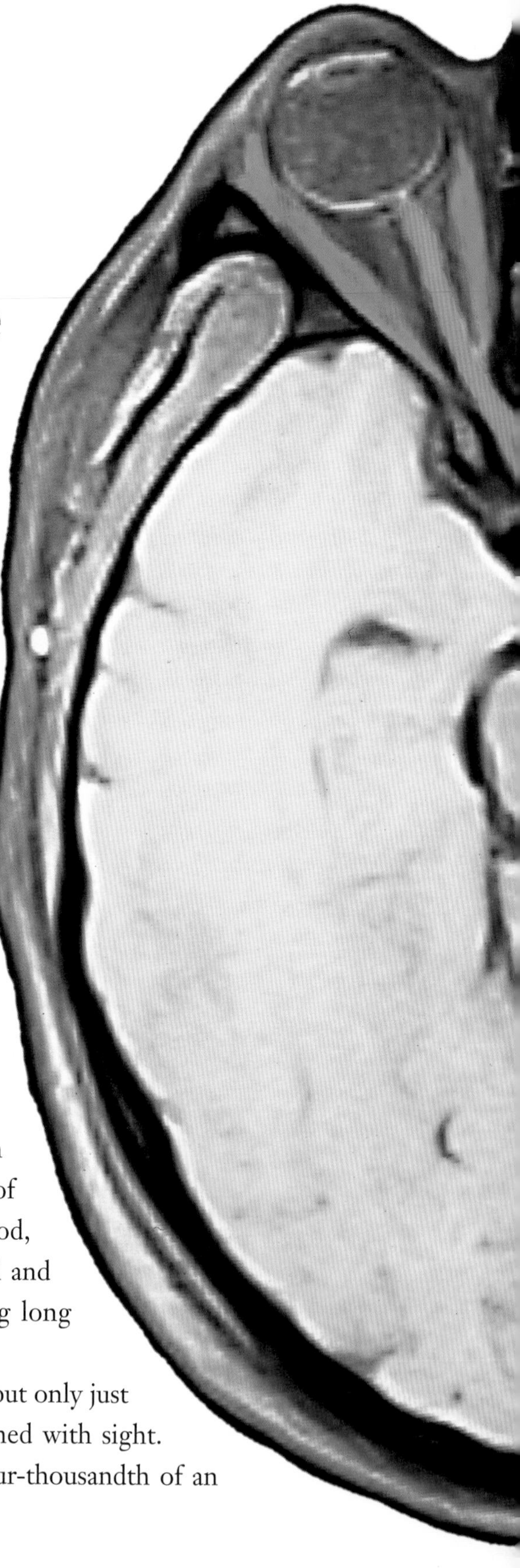

Vision
– the dominant sense

Humans have five senses, or more if the sense of touch is sub-divided. Pressure, pain, heat, cold and touch are each looked after by distinct receptors, with most for pain and fewest for cold. Many animals have much keener senses informing them of their outside world, with reality (as neuroscientist Colin Blakemore phrased it) being 'that which is biologically necessary for a particular animal to detect'. For some odours a dog's detection is one million times more sensitive than a human's. Bats and dolphins are supreme at echo-location. Prawns know depth to within a centimetre. Owls see mice at night. Birds can accurately fly the oceans. Ticks can detect the warmth of nearby mammals. Some moths can smell each other over a mile away. Salmon find the same river years after leaving it. Certain invertebrates can detect nuclear radiation. Human beings, by comparison, are blundering lumps of insensitivity. If a man is walking through a wood, however stealthily, virtually every other mammal and bird will have seen, heard or smelt him coming long before he knows (if ever) of their existence.

Vision is the dominant human sense organ, but only just over 1% of the brain's cortex is actually concerned with sight. Each eyeball weighs 7 grams. Therefore one-four-thousandth of an adult's weight is eye (as against one-four-hundredth of a baby's weight). The body grows 20-fold after birth, but the brain only enlarges just under four times and the eye just over three times. An eagle's eye is relatively larger than a human's, and an owl's eyes occupy one-third of its head, but a human's relatively tiny eyes are his or her major reception system for information about the outside world.

The inside of a human eye (seen through an ophthalmoscope) is the closest anyone can get to observing the living brain. About one million nerve fibres lead from each eye to the visual cortex of the occipital lobe. The connections between retina and cortex are not direct, with detours even before reaching the primary visual cortex, so called because the received impulses are passed on yet again to other cortical areas. With those two million fibres from both retinas leading to the hundred million neurones within the primary visual cortex, and with that cortex only the first stage in the brain's management of optical information, it is not surprising that the pathways are still only hazily understood.

Researchers have tested the visual perception not of the plethora of sights seen in normal

Left An MRI scan shows the eyes, the optic nerves leading from them (red) and the optic muscles (green).

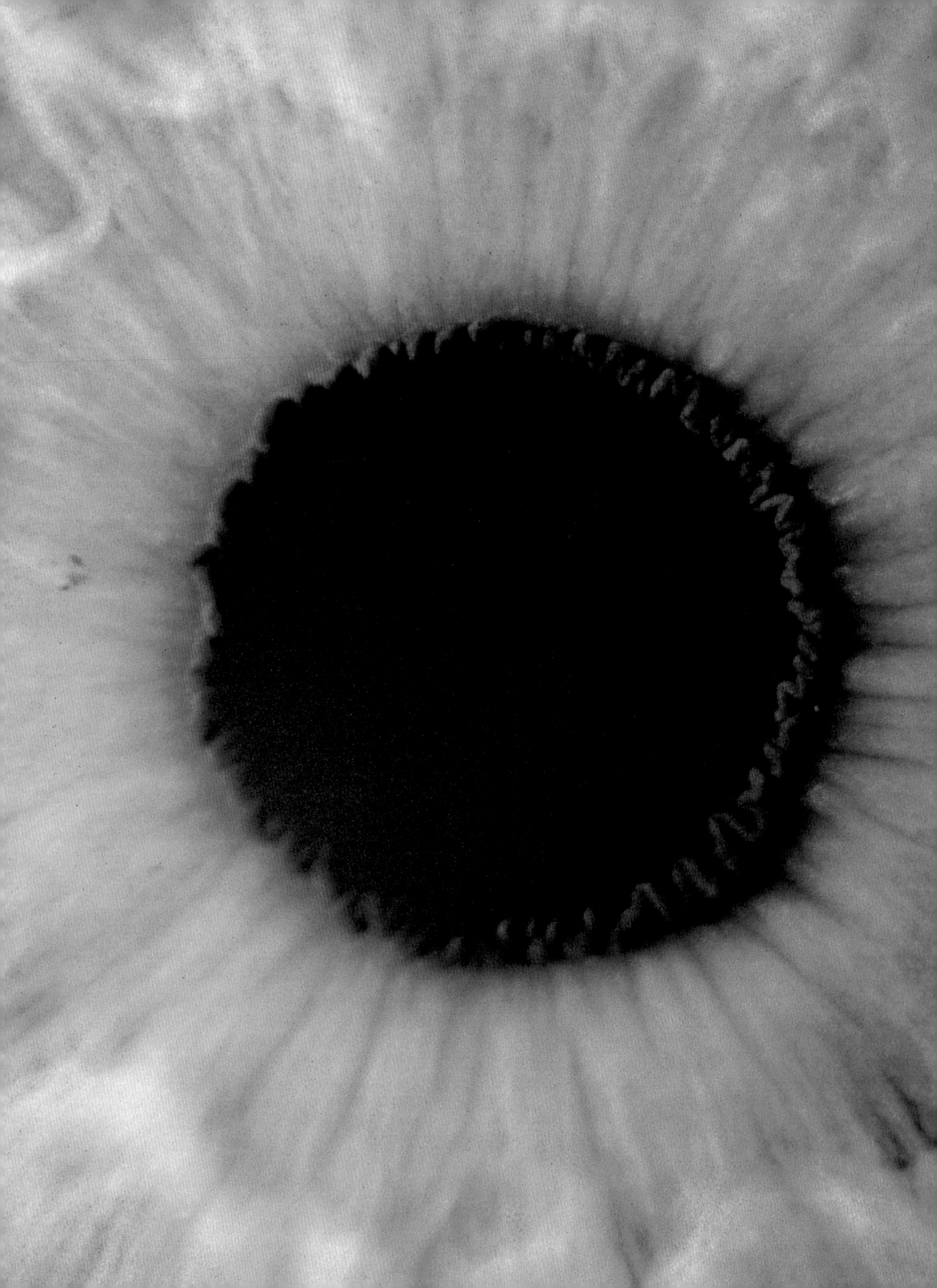

At the centre of the human eye (*above*), and surrounded by the iris, is the pupil (*left*). Light passes through the pupil and is focused by the lens to form an image on the retina. Each retina contains some 130 million rod cells (*overleaf*, coloured blue) which detect light intensity and transmit impulses via the optic nerve to the brain. The 6.5 million cone cells (*overleaf*, greeny-blue) do a similar job, but respond specifically to colour variation.

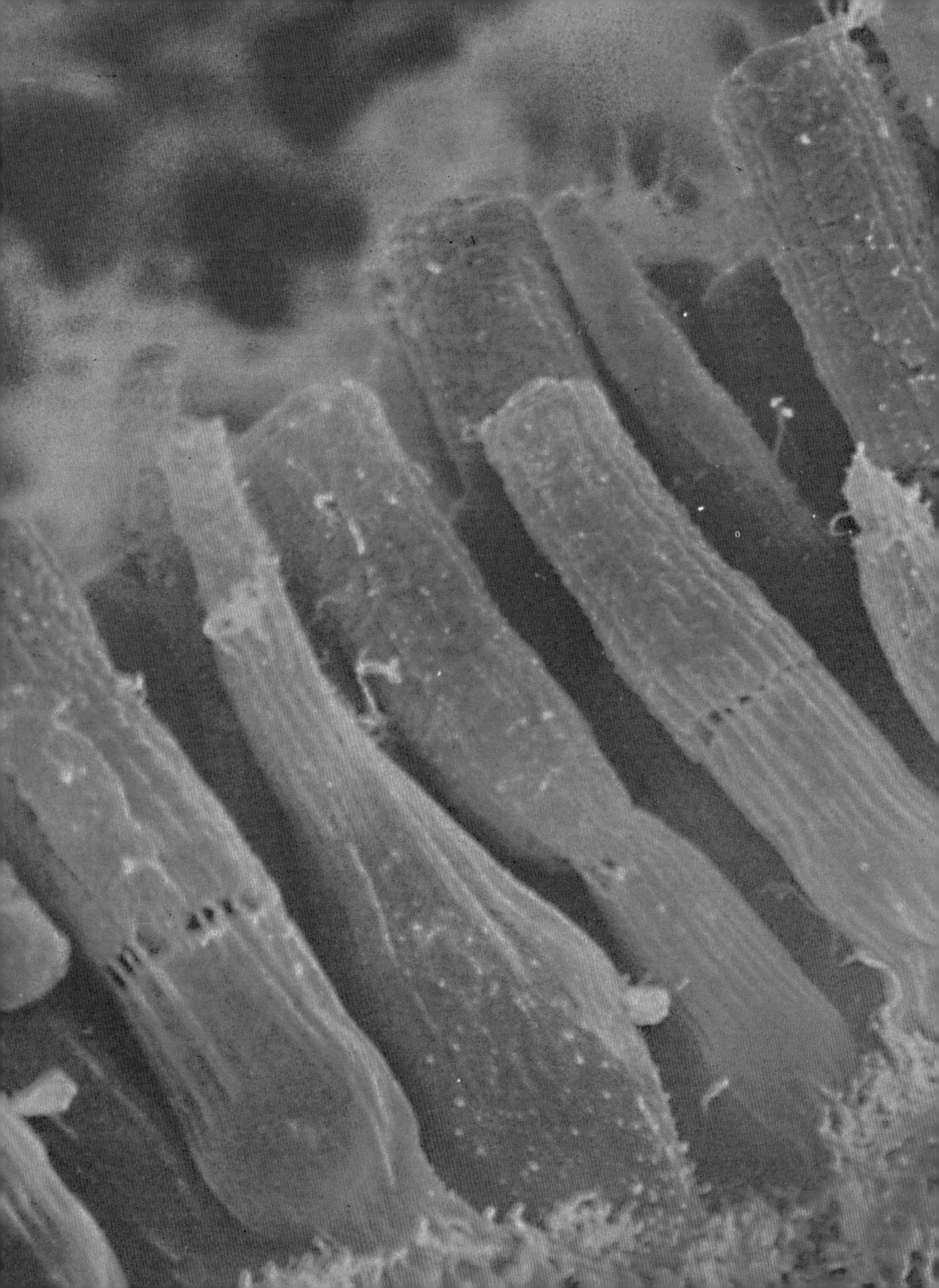

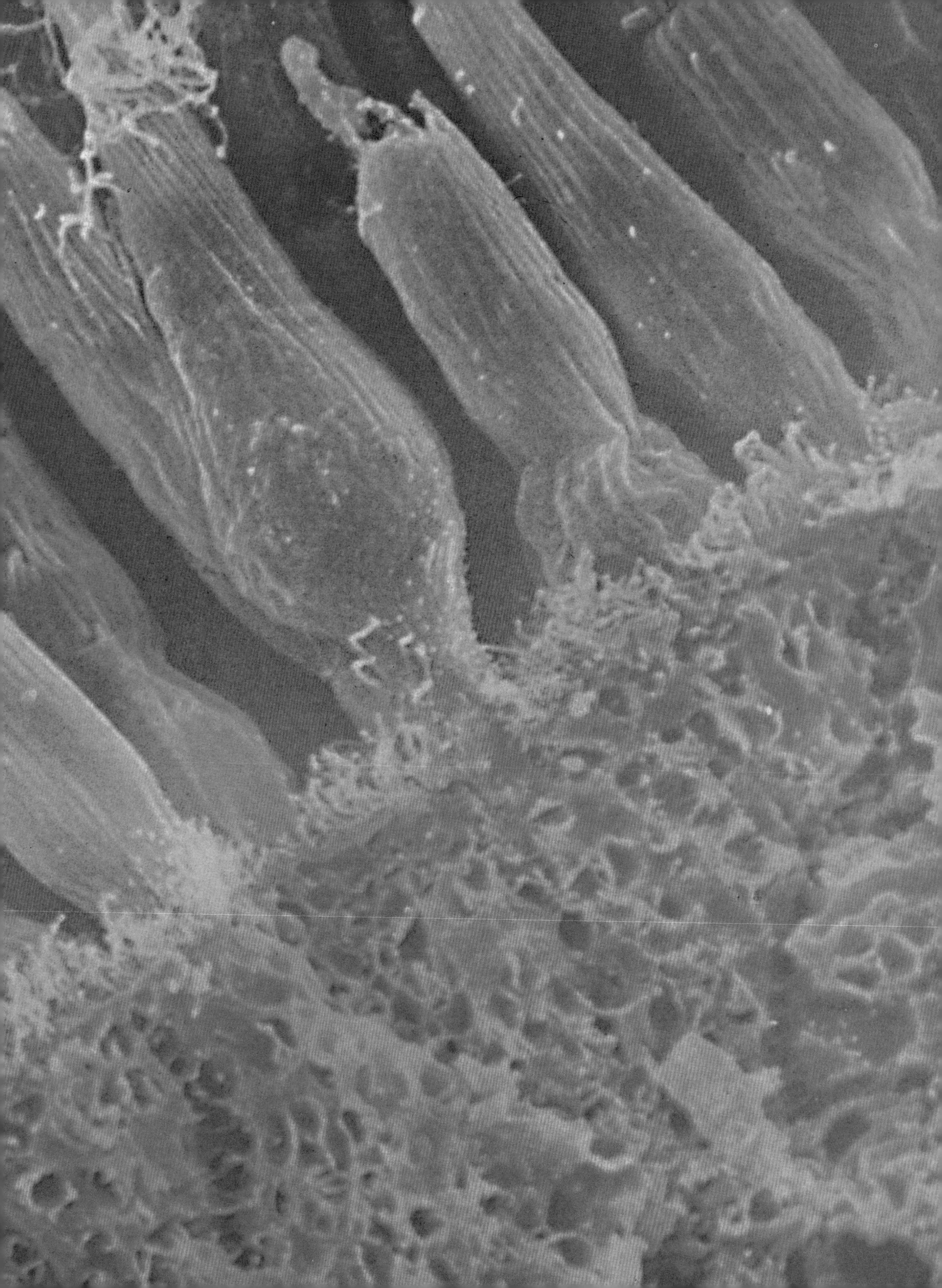

circumstances but of a straight white line on a black background. This particular input results in a set of regularly spaced patches of electrical activity within the cortex. The image is therefore scrambled before somehow being unscrambled, enabling us to perceive it as a line. To quote Colin Blakemore again: 'What does the brain do with the visual information after it has taken the image apart? If it only reassembles the features in some way to recreate the image, then why did it take it apart in the first place?' In many textbooks it is written that we see with our eyes. We do nothing of the sort. We see with our brains. We can even see with our brains without any visual input, as in dreams and hallucinations.

The retina at the back of the eye is first to receive the images, focused mainly by the cornea and partly by the lens. A man standing 3 metres from another is shrunk to an inverted image 10 millimetres tall in the other man's retina. The much smaller letters on an eye-chart subtend an angle of one-twelfth of a degree from the observer's viewpoint. On the retina these letters are only 0.025 mm tall, and they must be deciphered accurately for the observer to be judged visually competent.

Damage to certain parts of the brain can produce bizarre visual distortions, such as an inability to detect movement. Studying such conditions is helping researchers to map out the different brain functions involved in the visual process. Collecting the data from the light falling on the retina is only the beginning – a computer with a video camera can be made to 'see' in this way. But creating an 'intelligent' computer eye with even a fraction of human capability is proving extremely difficult. The goal of vision is to identify objects so that we can react to them in the appropriate way. More subtly, we must differentiate instantaneously between one human face, consisting of eyes, nose and mouth, and another one that shares the same features. Our brains, and those of all animals to varying degrees, have learnt to coordinate huge amounts of data on colour, shape and movement, to turn patches of light falling on the eye into information that causes us to greet a friend, eat a fruit or perceive and evade danger.

Colour perception is an important factor in human sight, but defective colour vision is a common trait. As with the normality of right-handedness, which is coupled with left-handed exceptions, so colour vision is normal but partnered by a high proportion of so-called colour-blind individuals. It is easy to wonder, if normal is preferable or advantageous, why natural selection has not eliminated the variants to a greater degree. Most colour-blindness is associated with the X chromosome, explaining why men have a higher incidence than women (the inheritance of one defective X manifests itself unless cancelled out by a second, normal chromosome as usually happens in women).

The higher primates have good colour vision, perhaps to enable them to spot fruits among the leaves in the forest or jungle. Smells are broken up in the jungle environment, so vision may have taken on a greater importance. However, most mammals are poor at colour discrimination. Wave a green rag at a bull, and he will still behave in the customary manner.

Wave a green flag at many male humans and they will not know it from a red one (explaining why the motorist in front may be dithering before an unfamiliar traffic-light, but not why red and green were chosen as key colours in the world of transport). There are six forms of colour confusion (a better name), ranging from common to very rare, but all more frequent in men. They are commonest among European males (8%), less common among Asian males (5%), less again among African males (3%), and least common among Inuit men (1%). The most widespread types lead to the confusion of red, yellow and green, but yellow/orange and blue/green confusion also occur. A few rare individuals are totally monochromatic, seeing everything as one colour.

Perhaps there were previous advantages in poor colour discrimination which would explain the current high proportions. To be colour-confused means being less distracted by colour, distraction often being the basic function of colour in nature. A colour-confused entomologist can be more adept at seeing insects by looking for their shape while undiverted by colourful camouflage. Everyone is colour-blind when light is weak. Ability to see red is the first to go each evening and blue the last. Blue is first to return each morning.

Abilities and enigmas

The brain is our most distinct, extraordinary and exciting possession. Our brains are attempting to unravel their own workings, but there is enigma at every turn. The smallest nick in one area can have devastating effects, and yet a crowbar can go *through* a person's skull without killing him. One such victim subsequently walked from hospital 'with a slight limp'. Phineas Gage famously survived a 6-kilogram tamping iron (for compacting explosive) passing through his brain. He lost consciousness only briefly, and later became a drunkard, but he still had the wit to sell his skeleton, cash in advance, before his death 12 years later.

Phenomenal mental abilities can astonish the rest of us, making us wonder if there is something different about their practitioners' brains. One German musician read an unknown (to him) symphony once before conducting it from memory that evening. An Edinburgh mathematician was asked to divide 4 by 47. After slightly more than half a minute of giving numbers, and having reached 46 decimal places, he said he 'had arrived at the repeating point'. And so he had. Wolfgang Mozart reported that 'an entire new composition' would suddenly arise in his head. At some convenient and later time he would write it down. All of this, it should be remembered, is being achieved by a hunter-gatherer's mind.

Memory experts can confound all of us, particularly when we have forgotten a phone number and they are remembering the number mathematicians call 'pi' to 8000 places. They have recently been scientifically investigated to discover if their brains have any special features. The answer is that their brains are normal, but they have learnt techniques for remembering. For instance, they turn numbers into stories – four, five, six can become 'your nice bricks' – and all of us are good at remembering stories.

Whether humans conduct symphonies, perform mental arithmetic, compose or learn phone books by heart, their brains are similar. The strangeness is that these organs were developed for a life-style so different from the way so many of us live today – or perhaps only seemingly different but equally complex. A friend recently spent a day with a South American Indian living a traditional kind of life. The two of them started from one particular tree and, after hours of running, hunting, stalking and catching, were abruptly back at their starting tree. 'How did you know where the tree was?' puffed the exhausted friend. 'It was in the same place,' replied the Indian.

Consciousness is an even bigger mystery. We do not even know how to define it. 'It is a vital part of living which defies definition,' wrote W. Ritchie Russell. 'It is the perception of what passes in a man's mind,' wrote John Locke three centuries earlier. 'It is the most obvious and the most mysterious feature of our minds,' added Daniel C. Dennett, getting nowhere near a definition. 'It is the knowledge that a man has within himself of his own thoughts and actions,' wrote Jonathan Swift rather more succinctly in the 18th century. One more definition, from a modern dictionary of 'the mind, brain and behaviour', stated that 'consciousness is being aware of oneself as a distinct entity, separate from other people or things in one's environment. The awareness is probably present to a varying degree in the higher animals as well as man, and presumably a function of the complexity of the living brain and its integrative power.'

Our 1.4 kilograms of nervous tissue – not our biggest organ but certainly the most demanding, not the most interesting in appearance and yet the seat of our individuality – is proving by far the hardest to comprehend. We do not know why it became so enlarged or how it is so clever. Somehow its billions of cells and its trillions of connections operate so that we not only dub ourselves *Homo sapiens* ('wise man') but actually merit the name.

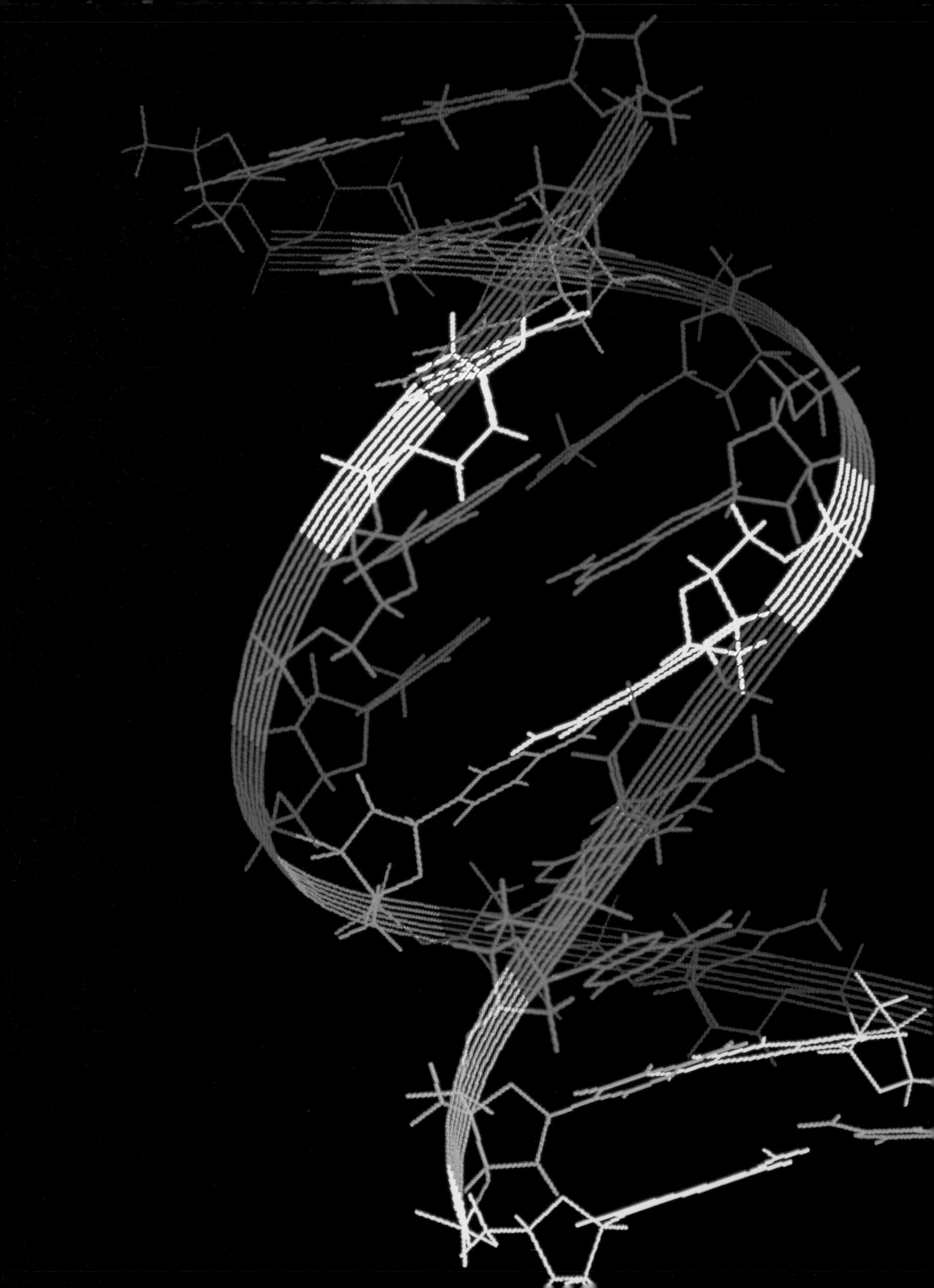

AGEING

All of us alive today are pinnacles, culminations of an evolution which led from invertebrates to vertebrates, from amphibians to reptiles to mammals, and then to the primates, of which Homo sapiens *is one kind. We are products of success. Our parents were successful in producing us, their parents were successful in producing them, and so on back to the beginning. Before such thinking goes to our heads we should remember that the same applies to every living thing. Being alive is a triumph, the end result of a formidable endeavour by countless ancestors. They all left genes to carry on their kind, and those genes, much amended through the aeons of time, have led to each and every one of us.*

As we journey through the early stages of life our bodies change and develop to meet the challenges of each new age. Year by year, from infancy to childhood to puberty, we are continually developing. On achieving adulthood – one more triumph in the long line of achievement – we are as strong and fast and tall, and as physically proficient, as we will ever be. We have also become sexually mature, and are then able to pass on our particular assortment of genes.

And then what? As all of us learn sooner or later, we age.

Inset The face of a man in his nineties shows all too clearly the wrinkling and sagging of old age, as well as white hair and the concentrations of melanin (skin pigment) known as liver spots.

Main picture A model of DNA or deoxyribonucleic acid, the molecule which carries our genetic inheritance. Antioxidants in certain fruit and vegetables help to reduce the chance of errors in the DNA which are generally found to increase with age.

JOTE AND VIOLA

Jote and Viola live on a farm in Kansas, USA. They have four children and six grandchildren and are about to celebrate their momentous 45th wedding anniversary. 'We've had an awful good life,' said Jote, 'but it's been hard.'

When she was 58 Viola was afflicted by spinal meningitis and, six months later, cancer. Both were treated and cured, but one year further on a different cancer assaulted her. Jote, 14 years her senior, also began to feel his age. 'We don't care to run around as much as we did. We're more likely to stay at home, watch TV, read.'

Both worry about Viola's cancer, and whether it might recur. But he, aged 76, continues with the farm while she, aged 62, still works mainly in the house. 'She gets tired quicker,' says Jote, 'but she's more tolerant than she was.' Viola agrees. 'We get wiser, definitely smarter, as we get older. But my eyes have gotten worse each year. Maybe it's the ageing process or the sewing that I do. But everyone worries about their health as they get older. So it's eyesight, and false teeth, and all those fun things.'

Their son, having arrived for the 45th anniversary, was surprised by the amount they had aged in the past couple of years, 'their hair a little greyer, their backs not quite so straight, a few more wrinkles and a general mellowing. Their aches and pains are bothering them a little more.'

On the subject of death Viola sounds almost matter-of-fact. 'Me, I don't worry about dying. That's just the next step in life. Life goes on after that, so I don't worry about the dying. That's out of my hands.' Jote wonders 'which one will be left, and the other gone?' He does not want Viola to have to work or 'go to an old folk's home'.

There is a single certainty for both of them. Old age is not only embracing them but making them think in a fashion most improbable when first meeting each other and raising their family. They have sown their seed, and ageing is now their lot.

From our pinnacle of physical achievement the path gradually turns downhill. We become progressively less capable in almost every way, particularly towards the end. Our appearance changes. We are likely to become shorter, more hunched, stiffer than formerly, grey haired, balder, more wrinkled, with drier, thinner and discoloured skin, with a double or protruding chin, a different voice, poor hearing, poor eyesight, varicose veins, inefficient memory, bigger ear lobes and a broader, longer nose. Death is not instant but we become increasingly vulnerable to it. After 40 the death rate doubles with every eight years in age, until all of us are caught by it.

We all know we face this fate, but why? Why, when evolution has adapted us so perfectly for surviving and reproducing, do we suddenly find things beginning to go wrong? Is it part of some grand design to make way for the next generation, or just a cruel accident?

Most researchers would now argue that ageing is not part of the evolutionary pattern at all, but completely outside it. Evolution is driven by natural selection, in which individuals whose genes give them some sort of advantage pass those genes on more successfully than their less genetically fortunate peers. The human race has never cultivated genes which endow health and strength in old age because it has never needed them. Evidence from the skeletal remains of ancient man, and in more recent times from historical records, suggests that for the overwhelming majority of human existence, life has been brutally short. Some individuals, and perhaps some societies, did live longer, as indicated by the Biblical reference to 'three score years and ten'. But mass old age as we know it is thought to be a new phenomenon, never witnessed by our race before.

The second argument against a genetic 'purpose' for ageing is that it occurs after the time when an individual can reproduce and pass on his or her genes. Genes that only give advantages late in life will not help us in our reproductive years, and therefore play no part in natural selection.

We are the pioneers of old age, and nothing in our genes prepares us for it. It is as though a sophisticated machine is no longer required by its owners. Deprived of maintenance it gradually becomes more unreliable, loses some of the functions it once had, and eventually grinds to a halt. So it seems to be with the body. Once they have seen us through development and reproduction, our genes abandon us to our fate.

Theories of ageing

Humanity has long sought to explain ageing, and the theories put forward have been many and varied. The 'rate of living' theory, popular at the beginning of the twentieth century, postulated that every animal was born with a finite number of heartbeats. If the heart beat fast, as with mice, life was short – something along the lines of 'live fast, die young'. If the heartbeat was slow, as for tortoises, life would be long. Unfortunately there turned out to be too many exceptions and the rule was disproved.

Science has not yet come up with a universally accepted theory of ageing, but with the increasingly rapid growth in our knowledge of molecular biology it is not surprising that the search has become concentrated at the cellular level. This may help explain one of the more puzzling questions of ageing. We know that most tissues in the body are in a constant state of turnover: the lining of the gut is replaced every three days, the skin is continually shedding dead cells as new ones push to the surface, even the skeleton changes its cellular make-up completely every seven years. Very little of our bodies is more than ten years old, so why don't we carry on looking (and functioning) as good as new? Why don't we live for ever?

Each cell carries the body's full genetic code in the DNA stored in its nucleus, and the DNA controls the cell's activities and responses. In chemical terms DNA is a huge, extremely complex molecule. Yet every time a cell divides the DNA must replicate itself to produce a perfect copy. To prevent errors – known as mutations – creeping into the genetic code, the DNA is constantly checked over or 'proof read' by enzymes within the cell nucleus. When they encounter a damaged section or a mistake in the code, the checking enzymes summon repair enzymes which cut out and replace the faulty portion. This amazing team of molecular maintenance engineers is efficient but not foolproof: some mutations go undetected and are passed on to the daughter cells when the cell divides. The damage repair system also becomes less effective as we grow older, so that the DNA being passed on to new cells contains more and more errors. As DNA is the blueprint for protein production, this leads to defects in the proteins that make up the cell's structure and control its functioning. In particular, cells can accumulate mutations that cause them to grow and divide uncontrollably, the condition we know as cancer.

● *Right* Two daughter cells formed from mitosis, or cell division. Every time a cell divides, the DNA must replicate itself also. The result is two daughter cells with identical DNA. Occasionally errors are introduced into the DNA and these can result in cancerous cells.

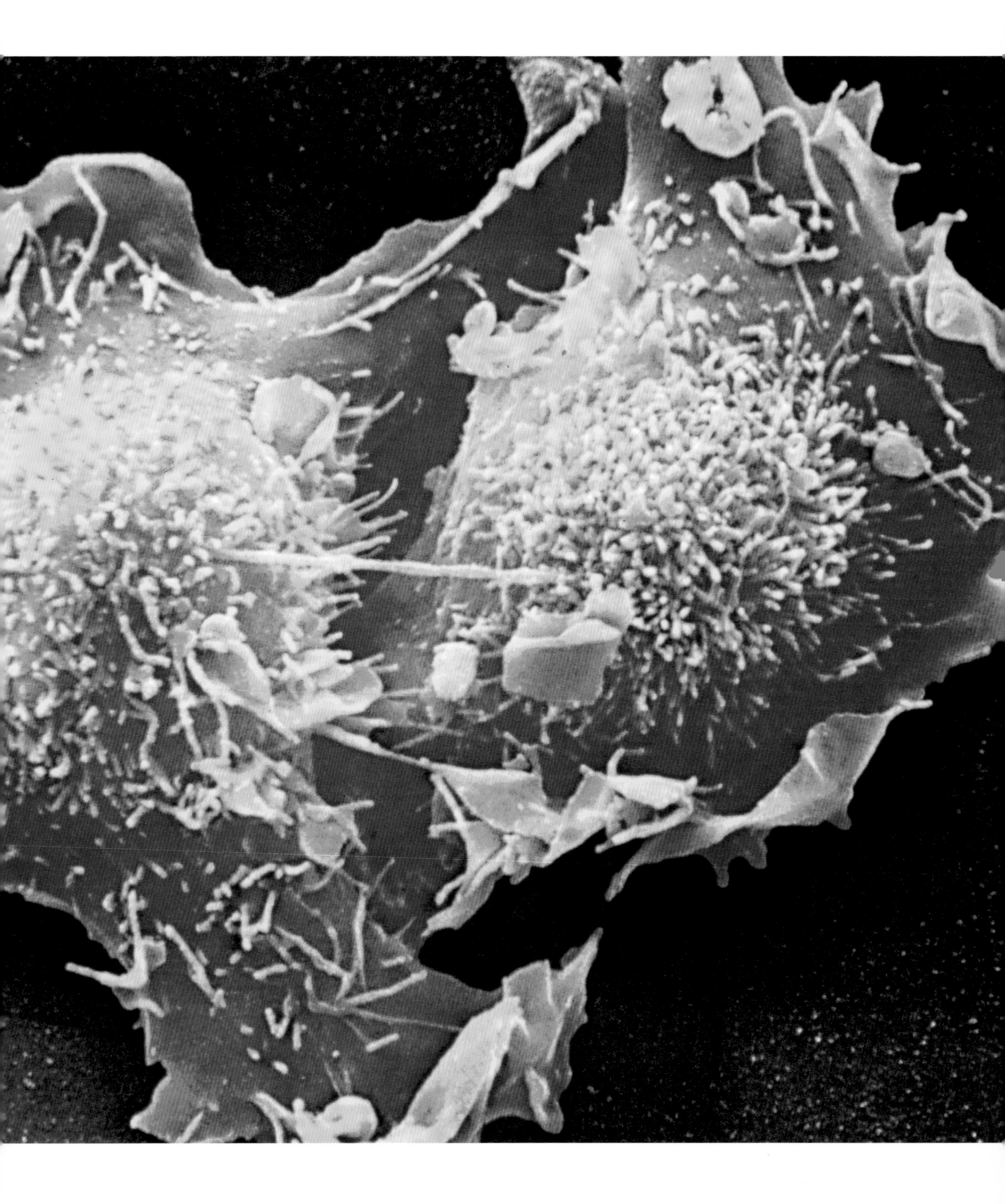

One of the major causes of DNA damage is free radicals, highly reactive molecules produced during normal metabolic reactions involving oxygen. Radicals will damage any other molecule they touch, and would wreak havoc in the cell if unchecked. To neutralise them the cell contains various compounds known as antioxidants, among them the vitamins C and E and various chemicals found in fruits and vegetables. Nevertheless, free radical damage does occur and is thought to be one of the mechanisms of ageing. It has been found that humans, the longest-lived of all animals, have much higher levels of antioxidants than shorter-lived species. One scientist has bred a strain of roundworms which have a mutant gene causing abnormally high production of antioxidants and live up to 70% longer than the original strain. Humans with a high dietary intake of fruit and vegetables (and therefore antioxidants) have a reduced risk of cancer, though this effect has not yet been reproduced with artificial antioxidant supplements.

The ageing brain

Free radicals play an important role in the ageing of the brain. Post-mortem studies have shown that the brains of older people have a higher content of proteins damaged by free radicals than those of younger people. Studies in gerbils revealed that this damage was accompanied by decreased levels of certain enzymes and by reduced ability to navigate mazes, a common way of testing animal brain function. However, when the older gerbils were given injections of an antioxidant to neutralise free radicals, the number of damaged proteins in their brains declined and their maze-running skills improved to match those of young animals. When the injections were stopped the protein damage returned and performance in the maze dropped back. This sounds like a potential elixir for youth, but long-term experiments have shown that cells receiving extra antioxidant eventually cut back their own natural production, leaving no net increase.

The ageing brain also undergoes changes in structure. It is one of the few parts of the body where cells are not replaced, so the number of neurones or nerve cells decreases as some atrophy and die. Even in healthy people there is a loss of cells in areas of the brain important for learning, memory and complex thought processes. Neurones that remain may conduct nerve signals less effectively because free radicals have damaged the fats in their cell membranes. However, there is evidence that some may actually grow new branches to compensate for loss or shrinkage elsewhere, and this capacity for regrowth could perhaps one day be stimulated with drugs.

Although these changes sound frightening and depressing, it is encouraging to note that for most people they have little impact on the complex web of interactions that we call the mind. The gradual neuronal loss does not automatically affect our practical ability to think or reason. Tests on healthy people in their seventies and eighties show only slight reductions in performance of memory, perception and language, provided that people are given enough time and do not feel anxious about the tests. However, nearly everyone will admit to some degree of memory loss towards the end of life. People frequently claim to remember 'the old days' as though they were yesterday, while yesterday itself is already hazy. In reality the young can remember childhood better than the elderly, but the elderly remember the present less well.

Memory has three components: registration, retention and recall. The memory loss of old age is largely a problem of recall, as a common memory game can illustrate. Twenty different objects on a tray are examined for one minute and then removed, leaving the contestants to remember them. Old people score less well at this than youngsters. However, if 20 more objects are placed on the tray and the players are asked to pick out the original 20, old people will do just as well as young. They can register and retain information without difficulty, but are let down by their ability to recall it. Problem-solving and learning abilities also deteriorate, but often not until very late in life.

Just as muscles become weaker from inactivity, brain function deteriorates when unused – a case of 'use it or lose it'. Rats in barren cages have fewer neuronal connections. Humans whose brains develop more connections

during life have less risk of memory loss, but the quality of blood and therefore oxygen supply to the brain is also very important. As one physician wrote: 'if you have four good tubes going to the brain – two carotid arteries and two vertebrals – two good tubes going to the kidneys and some good coronary ones, you can be an active, intelligent and aggressive man at the age of 80, whereas if they are defective you can be an enfeebled old dotard at the age of 50.'

What no one pretends is that brains improve with old age. They may become more skilled at cutting corners, seeing the wood for the trees, disregarding irrelevance, but this is not improvement of the type witnessed almost daily in the very young. To be senescent, and then senile, is to realise the brain's decreasing range of skills, and to grow 'both wiser and sillier' as the Duke of Rochefoucauld phrased it. 'Senile' is an emotive word and much misused. It means 'old', not demented (literally 'out of mind'). Senility and dementia may go together, but not necessarily. Madame Jeanne Calment of France, who died in 1997 at the age of 122 and was therefore certainly senile, was examined over a six-month period but showed no evidence of dementia. The ventricles and fissures in her brain were enlarged and there was atrophy of her lobes, all indicating shrinkage. But she was certainly mentally intact, even living independently until the age of 115.

Dementia, the severe and progressive loss of mental faculties, is perhaps the most feared aspect of growing old. In fact, most people escape dementia, even in extreme old age. One study put its prevalence at about 0.7% at age 60–64, 1.4% at 65–69, 3% at 70–74, 6% at 75–79, 11% at 80–84, 21% at 85–89 and 39% at 90–94. Other reports put the prevalence at 85+ anywhere between 15% and 50%.

Whereas normal age-related brain changes can lead to mild loss of memory or slowing of thought, dementia is thought to be caused by specific factors that lead to particular and excessive alterations to the brain. Thus dementia is a disease, not a normal consequence of ageing.

● *Left* Madame Calment, of Arles, Southern France, lived longer than any other human being ever documented.

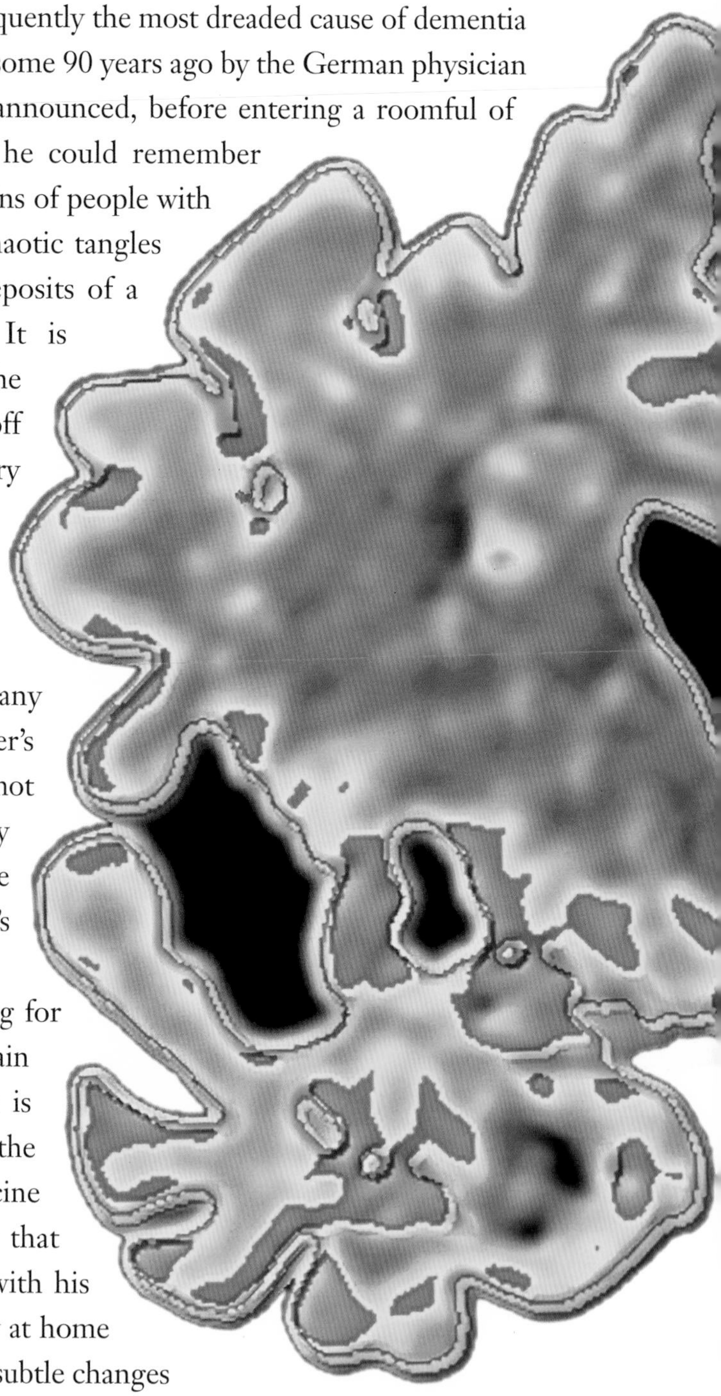

By far the commonest and consequently the most dreaded cause of dementia is Alzheimer's disease, first described some 90 years ago by the German physician Alois Alzheimer. (A friend recently announced, before entering a roomful of acquaintances, that the only name he could remember these days was Alzheimer's.) The brains of people with Alzheimer's become clogged with chaotic tangles of protein fibres and plaque-like deposits of a protein known as beta amyloid. It is thought that these plaques damage the cells around them and gradually cut off areas of the brain involved in memory and thinking. Areas controlling the general functioning of the body are not affected until very late in the disease, so that the sufferer is often physically healthy and capable of many more years of life. Why Alzheimer's happens in some elderly people but not others is not fully understood. A faulty gene has been discovered in some victims, but most cases of Alzheimer's are not thought to be inherited.

Alzheimer's disease is devastating for all involved, placing tremendous strain on carers and relatives. The onset is gradual and subtle. An article in the New England Journal of Medicine described the social impact, noting that 'when a demented person is living with his family, especially his wife, he can stay at home longer. In fact, adaptation to the first subtle changes of dementia is often automatic and unnoticed. The keeping of the cheque book changes hands. Grooming is inspected by another person. Food preparation is gradually modified so that the items are easier to eat, chew and swallow. Sandwiches may take the place of food requiring a knife, fork and spoon. In this way the spouse of a demented person may preserve that person's appearance of mental integrity long after it is gone. There may be a gradual shift in roles. A daughter may treat her demented mother more like her child. A formerly dependent wife may become the family manager and factotum. The breadwinner may also become cook, housekeeper and launderer.'

Left A vertical slice through the brain of an Alzheimer's sufferer (left) compared with a similar slice through the brain of a normal individual (right). The shrinkage is due to the degeneration and death of nerve cells.

Grey hair and baldness are the stereotypical badges of old age, but changes in the hair vary widely between individuals.

An average scalp has about 100,000 hairs, each growing from its own follicle. The number of follicles is fixed and is determined before birth. Long chains of the protein keratin are secreted at the bottom of the follicle and pushed out through the scalp, the visible hair being a dead, inert structure. Scalp hair in Caucasians grows at about a centimetre a month, or a third of a millimetre a day. If this total is multiplied by the number of follicles, a formidable 12 kilometres of hair is grown annually by the scalp alone.

Each follicle grows hair for two or three years and then becomes dormant, causing the hair to fall out. Later it will become active again to form a new hair. The cycle is staggered across the scalp, with about 10% of the follicles in the resting phase at any one time. The maximum length of any hair is determined by the length of the follicle's active period, as well as the rate of that hair's growth.

Adults, both male and female, lose about 70–100 hairs a day, and loss can be increased by a variety of factors including diet, illness and stress. In youth there is a balance between loss and replacement, but as we age the loss speeds up without a corresponding increase in new growth. The resting phase of the follicles lengthens, and some are switched off altogether by

● *Above left* Two shafts of scalp hair. They are covered in scaly plates of the protein keratin, and this structure is thought to stop them from matting together.

● *Left* Male pattern baldness. The hair loss is a hereditary response to the hormone testosterone.

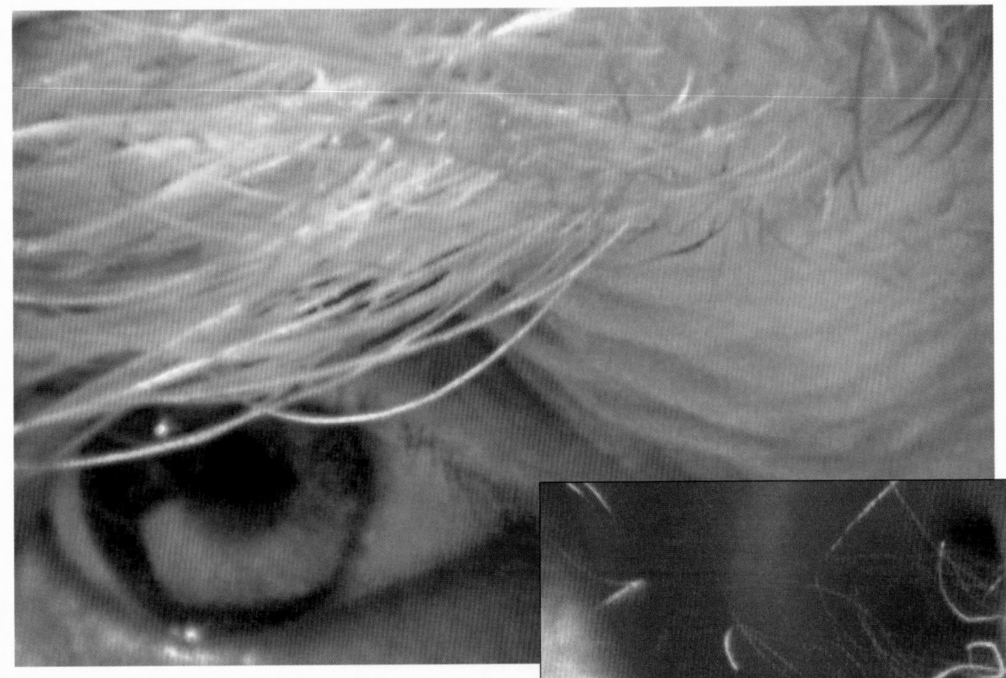

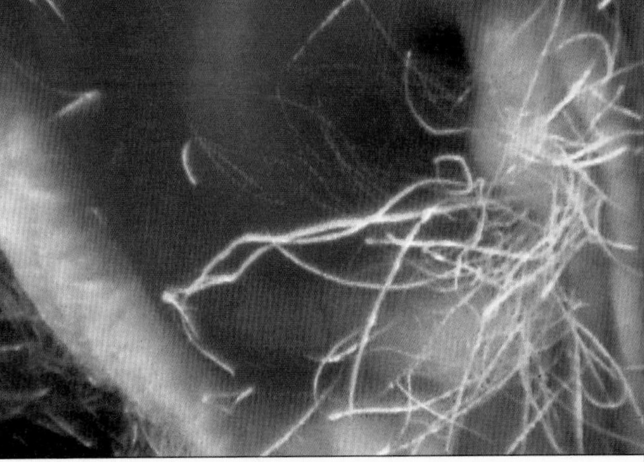

Eyebrow hair (*top*) and ear hair (*below*) increase in old age, an opposite effect to the thinning of scalp hair.

deactivation or destruction of the cells that make the protein shaft. This happens in both men and women, and is thought to be under genetic control. As a result, the hair thins.

Men are also prone to 'male pattern baldness', which has a different cause from age-related thinning and produces the characteristic patch pattern. Instead of being destroyed, the follicles stop producing normal hair and begin to make delicate structures called vellus hairs. These reach only short lengths and are so fine that they are difficult to see with the naked eye, thus making the head appear bald. Eventually the hair follicle dies altogether. This type of baldness has a strong hereditary component: a young man can get a good idea of whether he will go bald by looking at older male relatives. Some researchers believe the gene is passed down the female line on the X chromosome, but it normally lies dormant in women because it is triggered by high levels of androgens, the male hormones. In men who do not carry the gene the hormones have no effect on their hair. Where the gene is present, the only way to avoid baldness is castration before puberty!

Paradoxically, men find some other types of hair become more vigorous as they get older. The beard grows more luxuriantly, the eyebrows become bushier and hair in the nose and ears becomes more noticeable. This is also due to androgens, but the follicles are programmed to respond to them in a different way.

Greying of the hair is also genetically controlled. Caucasians tend to start getting a few white hairs at between 30 and 40 years of age, and most individuals have some grey by 60. Hair is coloured because cells in the follicle secrete the pigment melanin as the hair shaft is produced. With age the number of pigment cells reduces until the hair is left unpigmented and therefore white. The grey appearance comes from white hairs being mixed in with pigmented ones.

Nobody knows why the pigment cells die out. On the bright side, unpigmented hair is wirier and coarser than coloured hair, and may fall out less easily. This could account for the theory that men who go grey early will be spared from balding. Contrary to popular belief, it is not possible to go white overnight with shock, though shock might cause coloured hairs to fall out more quickly, leaving a greater density of white ones behind.

The ageing body

Damage from free radicals and accumulated DNA errors play a part in ageing, but they are not the whole story. A powerful factor is simple wear and tear from the constant stream of insults bombarding our bodies, both externally and internally. Pollutants, toxins, natural radiation, viruses, bacteria and ultraviolet radiation from sunlight all assail us from the external environment, while internally our cells must cope with damaging waste products and the hormonal effects of physical and psychological stress. As we get older not only does wear and tear accumulate but our defence mechanisms, so effective in youth, begin to falter, perhaps because they themselves are damaged. Importantly, the functioning of the immune system is affected, making us more susceptible to diseases and cancers and occasionally causing the body to attack itself – a phenomenon called auto-immune disease. Rheumatoid arthritis is one of the increasing number of diseases known to be caused in this way.

'Old age does not come alone,' wrote Plato, who, by living to 80, must have known a thing or two about the process. Everyone has a different experience of ageing, but everyone who ages will encounter at least some of the bodily changes that are characteristic of advancing years. The changes in appearance are well known, and the processes behind them are explored in some of the boxes on the following pages. The functional changes are, if anything, even less welcome.

Sight and sound

For many people, one of the first signs of encroaching middle age is long-sightedness. It gets more and more difficult to read close up, until the book or newspaper is eventually squinted at from arm's length and there is no option but to get fitted out with reading glasses. This loss of close-up vision runs counter to what we might expect: surely if the eye is growing 'weaker' with age it should lose the ability to see far-away objects, not near ones? In fact, focusing on a near object places the greater demand on the eye: the lens must be squeezed by the eye muscles into a rounder, thicker shape in order to focus light rays originating close to it. Rays from distant objects require less bending and the lens takes on a more relaxed shape. As we age, the lens becomes less flexible, and also larger. It is one of the few parts of the body that carries on growing throughout life, and as it enlarges and stiffens it becomes more and more difficult for the surrounding muscles to squeeze into shape.

It also becomes harder to open the iris, the diaphragm that controls the size of the pupil and therefore the amount of light entering the eye. As the muscles become sluggish the eye responds less rapidly to changes in light, and the iris does not open as far, making it more difficult to see in poor light. This problem is compounded by changes to the lens itself, which becomes yellower and less transparent as the proteins that make it up degenerate. A 60-year-old's eye lets through only half as much light as a young person's, and by the age of 80 this has fallen to a quarter. In severe cases the lens develops a cataract, becoming so cloudy that the eye eventually goes blind. It has been estimated that more than two-thirds of the over-60s have some degree of cataract.

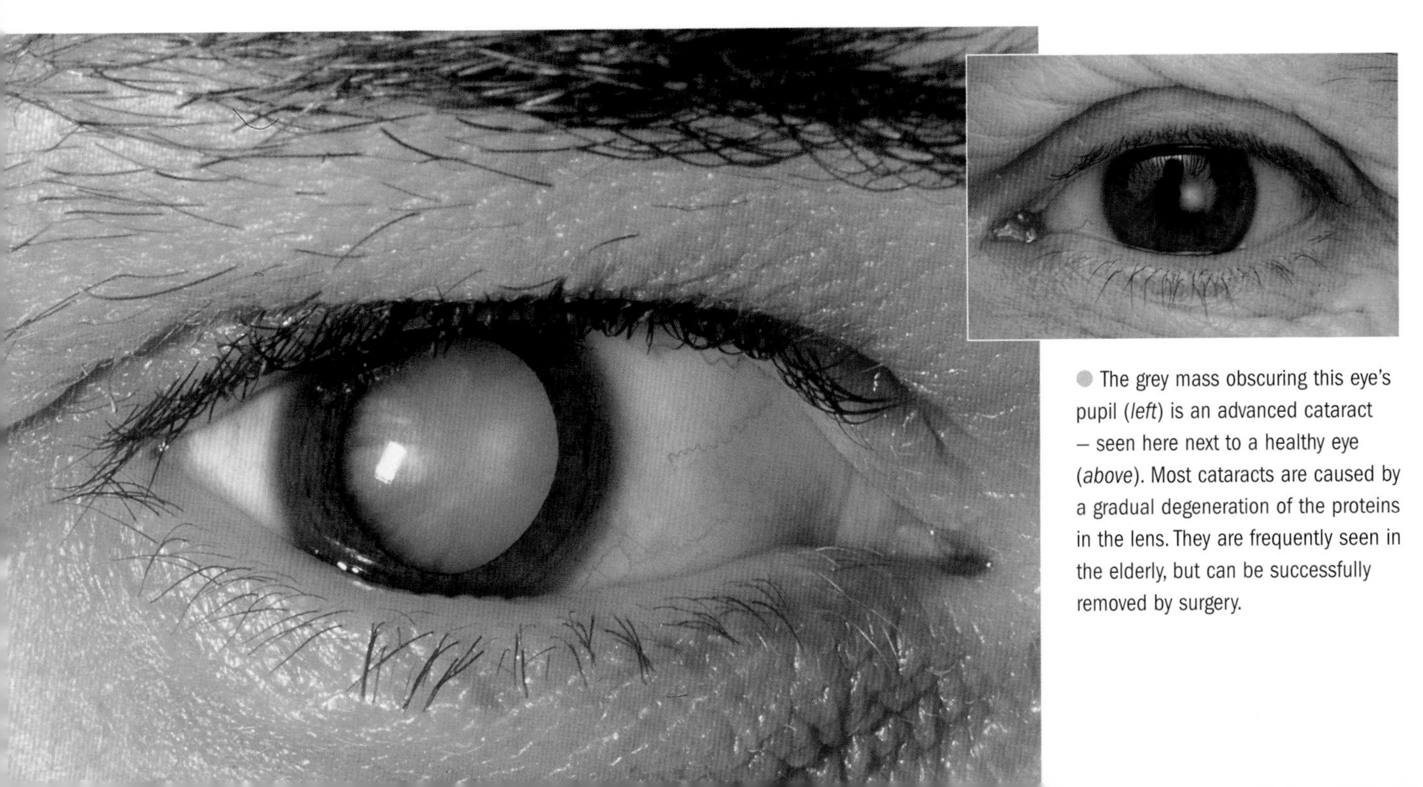

The grey mass obscuring this eye's pupil (*left*) is an advanced cataract – seen here next to a healthy eye (*above*). Most cataracts are caused by a gradual degeneration of the proteins in the lens. They are frequently seen in the elderly, but can be successfully removed by surgery.

THE AGEING SKIN

The skin is the body's first line of defence against the outside world, fighting a life-long battle against invaders such as bacteria and fungi, and against the elements. It is therefore not surprising that it begins to show wear and tear as we age. The appearance of the skin is one of the key factors we use in making judgements about how old people are, and chief among the attributes of ageing is the extent to which the skin is wrinkled.

Wrinkling is caused by ultraviolet radiation from sunlight. The more ultra-violet the skin receives over a lifetime, the more it will wrinkle – explaining the predominance of wrinkling on the face and the backs of the hands. Skin is naturally supple and elastic thanks to the protein collagen that is found in the upper layers. Pinch a piece of young skin and it snaps quickly back into place. Ultraviolet radiation attacks and destroys the collagen fibres, which instead of being renewed are gradually replaced by an abnormal form of elastin, another skin protein. The abnormal elastin is yellowish in colour, and with advancing age it begins to show through the thinning skin. Gradually the elasticity of the skin is lost, and it can no longer hold its shape against the downward pull of gravity. It begins to wrinkle and sag.

● *Below* Concentrations of melanin (skin pigment) known as liver spots on ageing hands.

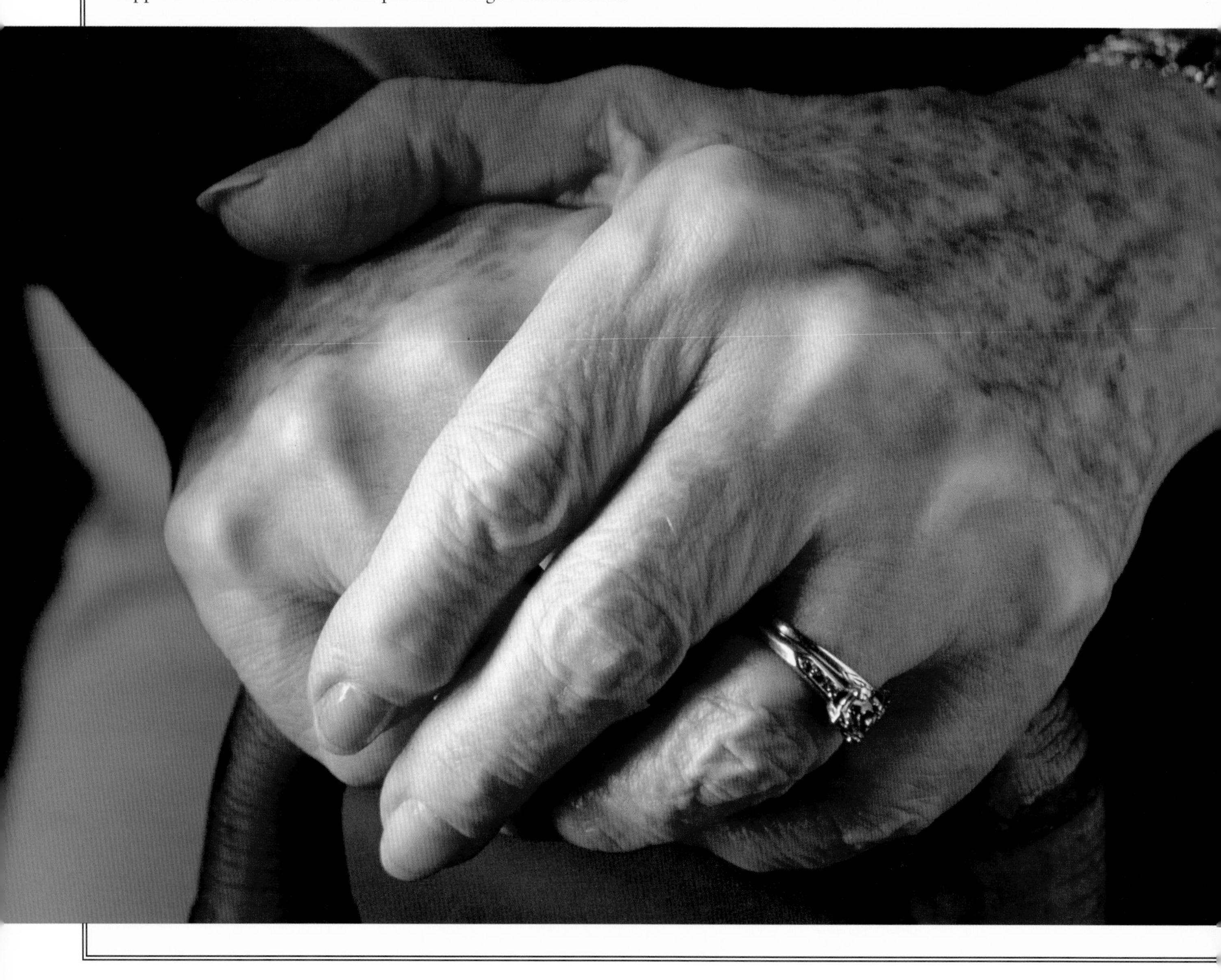

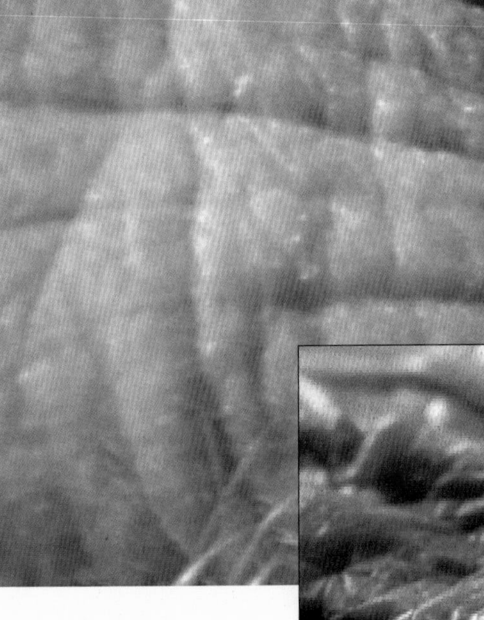

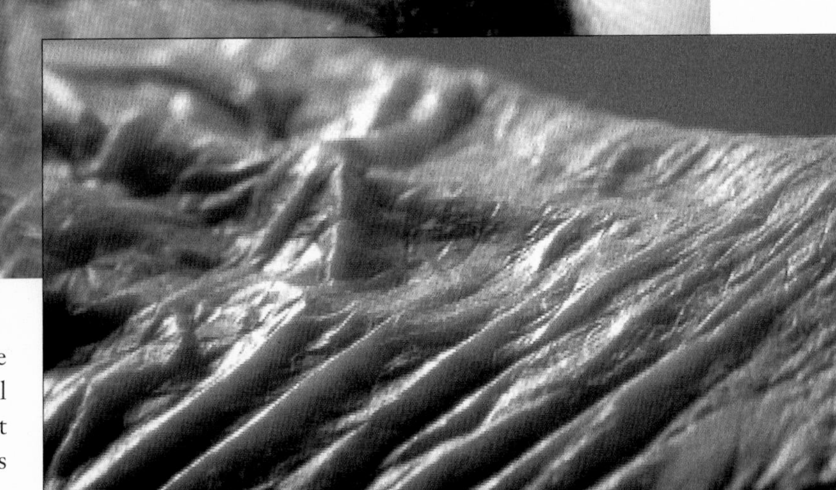

These close-ups of wrinkled skin on both the face (*top*) and hands of an elderly man (*below*) demonstrate how translucent and delicate the skin has become.

As the skin becomes less elastic, the repeated puckering caused by facial expressions begins to etch permanent lines. It has been estimated that it takes 200,000 frowns to make a brow line!

Protecting the skin against ultraviolet is the pigment melanin, produced by special cells at the base of the epidermis. All races have the same number of pigment cells: it is the rate of production that dictates skin colour. Although pale skins increase their melanin output when exposed to the sun – hence that holiday tan – dark skins suffer less ultraviolet damage and therefore less wrinkling.

Sometimes the melanin-producing cells begin to clump together in old age, due in some way to the amount of ultraviolet they have received. This produces the characteristic brown patches often seen on the face or the back of the hands. The English call them liver spots, though there is no known connection with the liver. The French are more poetic, referring to them as *les medaillons de cimitière* – cemetery medals. The equivalent in youngsters is freckles. No child is born with them: rather, they are a protective response against sunlight.

Ultraviolet rays also damage the layer at the base of the epidermis where new cells are formed, and over time affect its ability to renew the outer, visible skin layers. The cells alter their shape and become smaller, so that the thickness of the skin is reduced and it becomes more translucent. Even in areas of skin that are rarely exposed to the sun, ageing slows down the rate at which new cells are created. The loss of collagen means that the blood vessels are less well supported, making elderly skin more susceptible to bruising.

Potentially more dangerous is the damage done by ultraviolet to the skin's immune system. The number of immune cells in the skin declines naturally with age, but where there has been sun damage the count falls drastically. This lowers the level of immune surveillance, so that if skin cells begin to turn cancerous (another effect of ultraviolet), they are less likely to be weeded out and skin cancer may develop. Fortunately, many of the skin cancers associated with ageing are relatively benign.

Despite all these changes, the skin keeps on protecting the body throughout life. Nobody has ever died from skin failure.

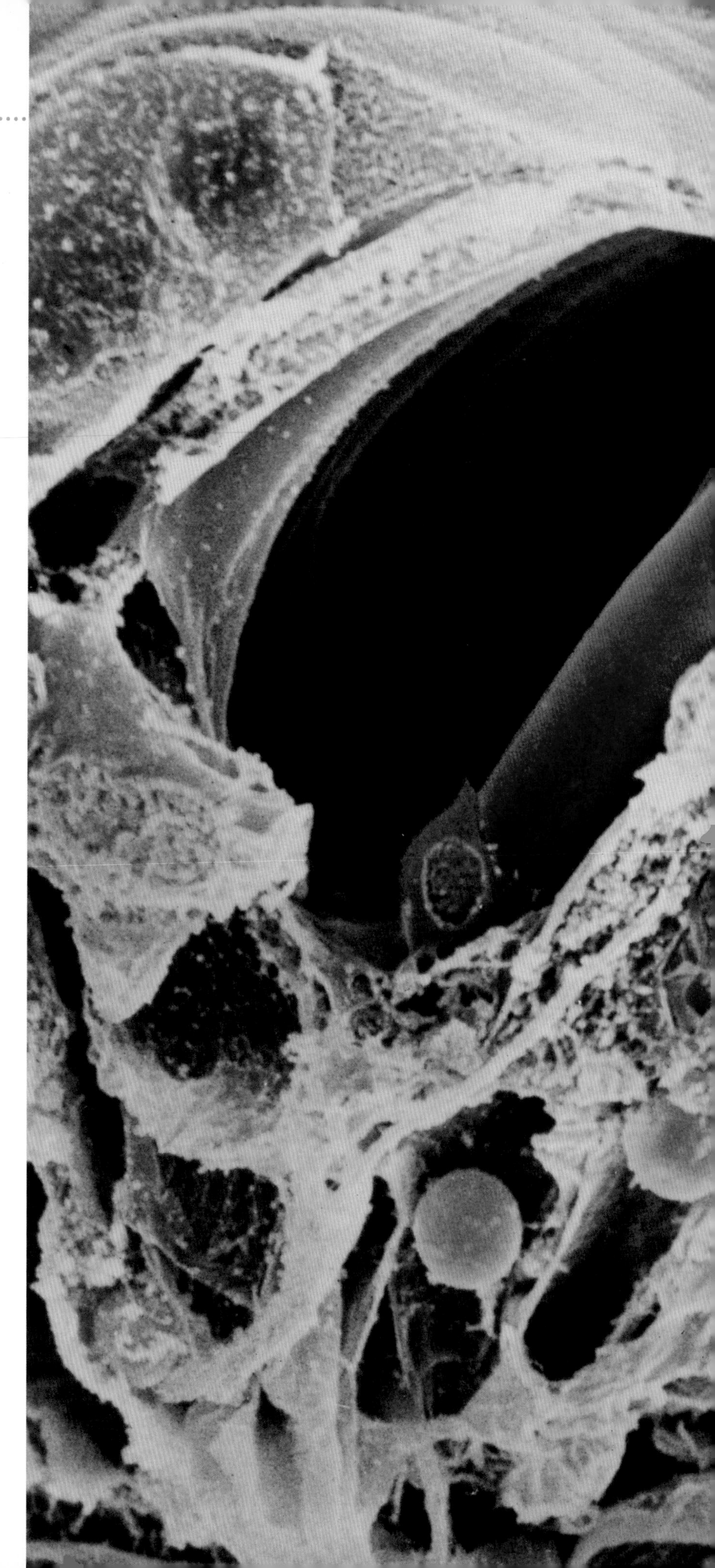

Right A section through the human inner ear of the organ of Corti. At the top right are four rows of hair cells, each containing about 100 hairs. The hairs translate sound vibrations into nerve signals, which are transmitted to the brain by the cochlear nerve. These cells die off in old age causing loss of hearing.

Hearing too is blunted by age. The loss of sensitivity begins from the moment of birth, so that by the age of ten we have heard more sounds than we will ever hear again. This rather depressing fact is caused by the gradual dying off of the exquisitely tiny hairs in the cochlea of the inner ear (see pages 174 and 175). Sound waves collected by the fleshy outer lobe pass down the ear canal and strike the cone-shaped membrane of the ear-drum. Movement of the drum in turn causes movement of the body's three smallest bones, minute structures known as the hammer, anvil and stirrup because of their shapes (see pages 16 and 17). The sizes of these bones are fixed when the ear is completed in the developing foetus, the stirrup remaining no larger than a grain of rice throughout life. The stirrup touches the oval window at the entrance to the inner ear and transfers the energy of the sound wave to the fluid in the cochlea, a spiral tube set within the skull bone that takes its name from the Latin word for snail.

Within the cochlea is the organ of Corti, where hair-like growths attached to nerve cells turn sound vibrations into electrical signals, much as a microphone does. Different cells are activated by different frequencies of sound, and send their signals to the brain to be decoded into the sounds we hear. The cells responding to high frequencies die off first, so we gradually become less sensitive to high-pitched sounds. The process is accelerated by repeated exposure to very loud noise: a sound can literally be deafening.

Because the eye and ear become less efficient, the quality of the sensory signals reaching the brain deteriorates, and the brain has to work harder to make sense of them. To an extent it will use the experience built up over a lifetime to fill in the gaps, but in old age we may be slower to interpret complex or unfamiliar environments.

The menopause

Of all ageing events to assault human beings, the female menopause is unique. It is not a slowing down or a deterioration but an ending. Although puberty results from hormonal changes prompted by the brain, menopause is caused by the ovaries running short of eggs. The average age of menopause in developed countries is about 51, with a normal age range of anywhere between 42 and 60. Neither the age of menarche (first menstruation) nor the number of children produced has any influence on the age of menopause. Once the 'change of life' has occurred the woman is no longer capable of producing children naturally.

All the eggs a woman will ever produce develop in the ovary of the foetus, which has around seven million potential eggs at mid-term. For some reason these egg cells then begin to die: by birth the number is down to one million, and by the time the woman reaches puberty she will have perhaps 40,000 left. The egg follicles in the ovary secrete the hormone oestrogen which drives the monthly menstrual cycle, preparing one or two eggs for fertilisation each month throughout the reproductive years. At the same time the store of unripe eggs in the ovary continues to die off, a process which accelerates after the age of 37 or so. Eventually there are not enough eggs left to produce the required amounts of oestrogen; men actually produce more oestrogen in their testes than women produce after menopause. The hormone progesterone, produced by the ovary in the second half of the menstrual cycle, is also lost. As hormone levels fall the length of time between menstruations increases, and eventually the periods stop altogether. Once the periods become irregular or seem to have stopped women may assume they are infertile, but this is not always the case. Menopausal women are occasionally shocked (or delighted) to find themselves pregnant when they had thought their reproductive years were over.

The disappearance of the sex hormones brings about physiological changes that often lead to unpleasant symptoms. Seventy per cent of women experience 'hot flushes', caused by instability in the tiny blood vessels of the skin.

AGEING BONES

Problems with the bones and joints are a major cause of discomfort and disability in old age. The two conditions responsible for most of this suffering are osteoporosis and osteoarthritis.

It may look inert, but bone is an active substance in a constant state of change. Adult bone is made up of 70% calcium and phosphate crystals and 30% protein, mostly in the form of collagen fibres. Dispersed throughout this scaffolding are living cells of two main types. The osteoblasts build up bone by making collagen and controlling calcium and phosphate deposition, while the osteoclasts resorb bone by attacking collagen and dissolving the mineral crystals. The reason resorption takes place is to regulate the concentration of minerals in the rest of the body, where any shortage would disrupt vital biochemical processes.

Up until the age of about 30, bone formation outweighs resorption and the bones become denser. The peak density reached in any individual depends 80% on genetic factors and 20% on environmental or life-style factors. A good calcium intake and plenty of exercise make for denser bone, while poor diet, lack of exercise, smoking and excessive alcohol can all reduce density. After age 35 there is a gradual loss of density, but bone loss and bone formation remain more or less in balance until old age, when the ability to make new bone is reduced.

In women bone density is influenced by the hormone oestrogen, and at the menopause the rate of bone resorption begins to increase. In some women the increase is very rapid, leading over time to severe weakening of the bones. This is known as osteoporosis. Seen through a microscope the effect is dramatic: the mineral scaffolding is eaten away and the bone becomes porous and brittle (see page 181). Fractures become more likely, especially in the wrist and hip. In severely affected women, even a violent sneeze can break a rib. The vertebrae of the spine sometimes crumble, leading to loss of height and curvature of the spine – the so-called 'dowager's hump'.

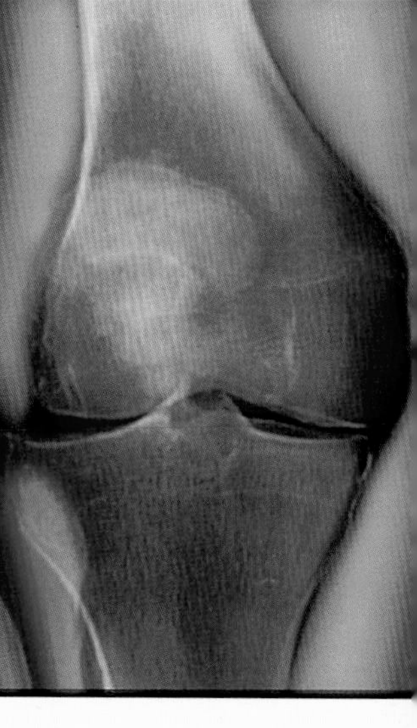

Nearly half of all women will experience some degree of osteoporosis, though hormone replacement therapy is increasingly being used as a successful preventative measure. The rate in men is about 1 in 12, partly because men's bones are denser to begin with and partly because they do not depend on oestrogen in the same way. Osteoporosis in men is thought to be linked to testosterone deficiency, at least in some cases.

Osteoarthritis affects the sexes more equally, with only a slightly raised prevalence in women. 'Arthritis' literally means 'inflammation of the joint'. In healthy joints the ends of the bones are protected by a smooth layer of cartilage, which acts as a shock absorber and allows for free movement. In osteoarthritis the cartilage becomes rough and worn, impeding the function of the joint. In an effort to compensate, the bone underneath thickens and spreads out at the sides. The membrane encasing the joint becomes irritated and secretes extra lubricating fluid, making the joint swell. The result is pain, stiffness and loss of movement, ranging from mild and occasional to severe and disabling. The most commonly affected areas are the knees, hands, hips, neck and big toes.

● A healthy knee joint (*right*), with the femur at the top articulating with the tibia below. When there is arthritis of the knee (*above*) some of the protective cartilage between the two bones is lost, causing a narrowing of the space between them. This frequently causes pain and a loss of mobility.

● Damaged joints can sometimes be replaced with artificial ones, hip replacements being particularly successful (*below*).

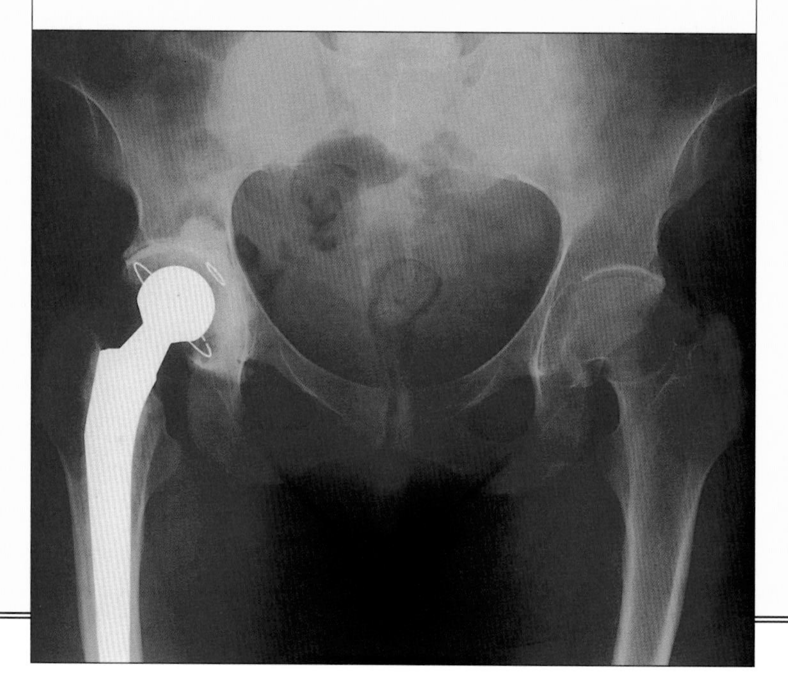

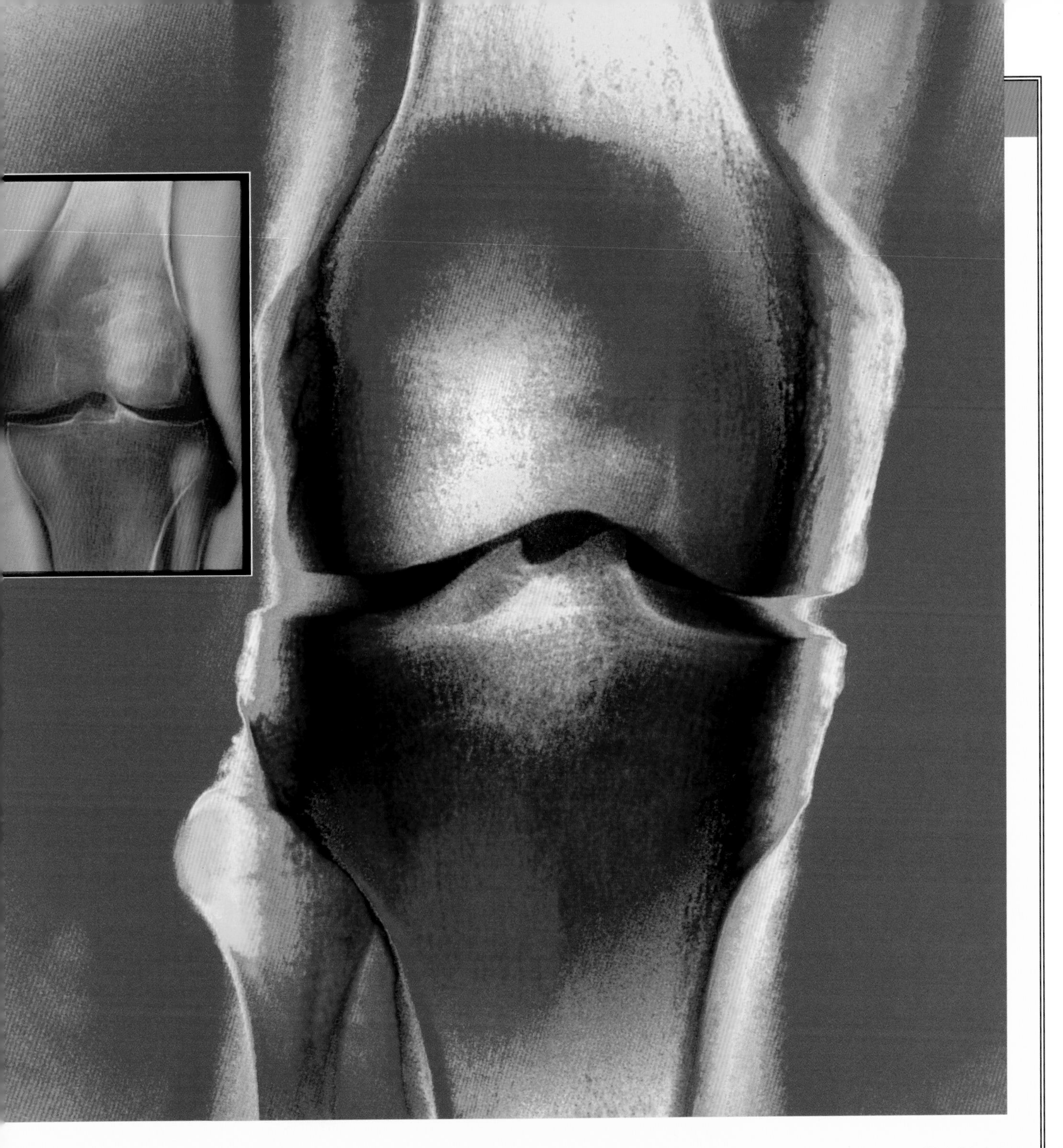

Arthritis used to be thought of as an inevitable consequence of a lifetime's wear and tear on the joint. New research suggests that the picture is not so simple. Ageing does play a part, but it is not a case of the joint simply 'wearing out'. The symptoms seem to reflect the joint's natural repair mechanisms at work, and many sufferers find the condition stabilises or even improves with time. The joint only begins to fail if the repair mechanisms can no longer keep up. In severe cases the best option is an artificial joint: hip replacements are very successful and the reliability of new knees is improving.

Some arthritis sufferers are credited with a strange ability to predict rain. This is not an old wives' tale: for some reason, the pain in affected joints actually does get worse when atmospheric pressure falls. This happens before wet weather, hence the uncanny forecasting knack.

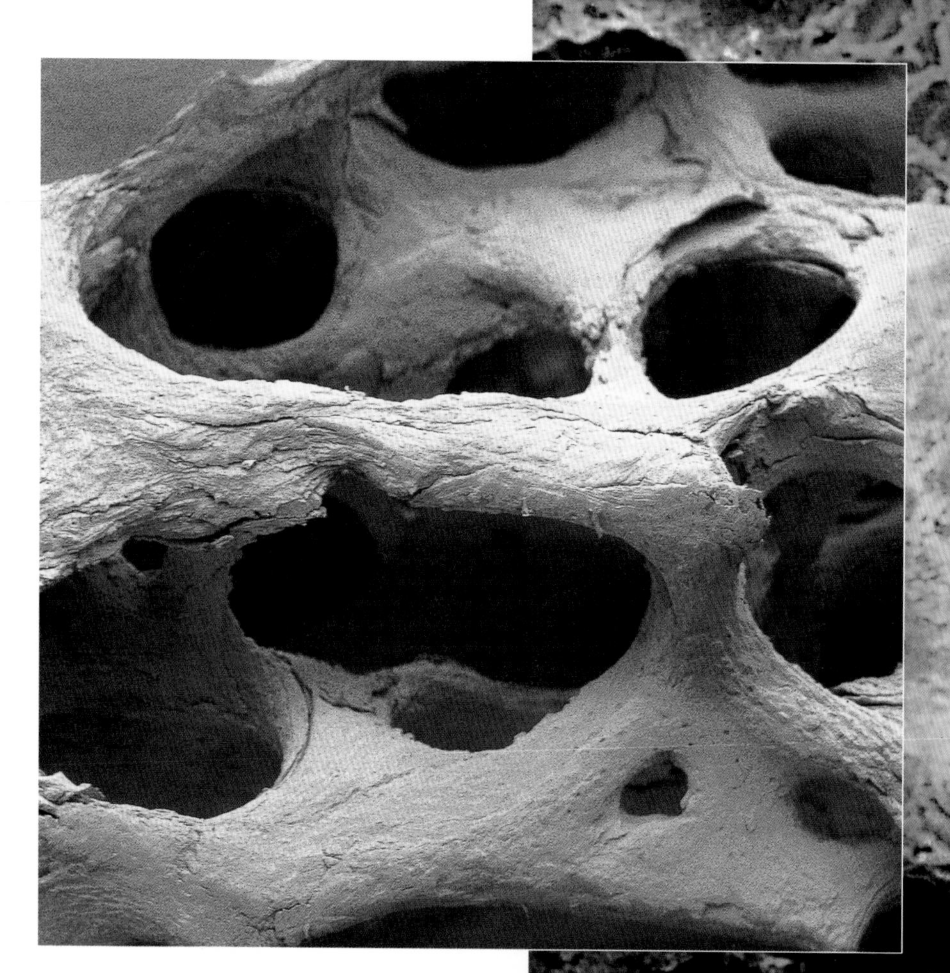

The image above shows healthy bone as seen through an electron microscope. The spongy appearance is normal for this type of bone. The bone on the right is affected by osteoporosis, and is clearly much less dense.

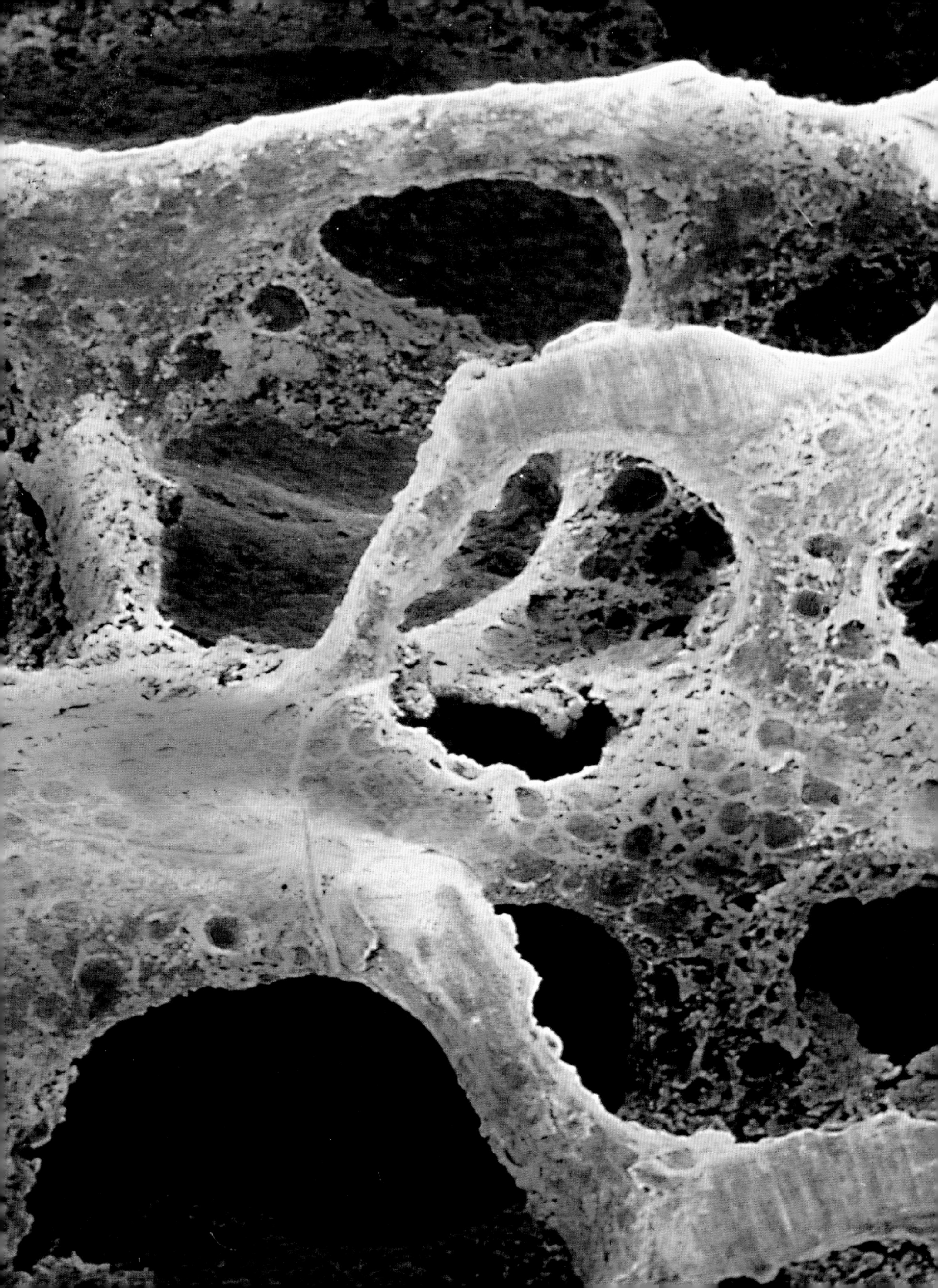

The menopause provides an interesting window into the poorly understood but increasingly acknowledged link between the hormonal system and the nervous system. Hormonal changes may lead to irritability, excessive tiredness, impaired concentration, depression or sleep problems, and in rare cases psychosis (severe mental disturbance). Sweating may increase, and some women suffer more frequent headaches. On the positive side, migraines almost always disappear. Sexual desire is not necessarily affected, but intercourse may be more difficult because the vagina generally becomes drier. The ovaries and Fallopian tubes shrink and the vagina shortens and narrows. The uterus hardens and shrinks to perhaps a quarter of its original size. The breasts lose some of their glandular material, either withering or replacing it with fat. Because the ovaries continue to secrete small amounts of the male hormone testosterone, certain male characteristics can become conspicuous, notably facial hair.

Menopausal symptoms may last for months or years, but sooner or later the body adapts to the changed hormone levels. However, menopause has come to be seen in some societies as a medical condition, almost a form of hormone deficiency disease like diabetes or thyroid problems. Increasingly, efforts are being made to counter its effects through hormone replacement therapy, or HRT (known as ERT in the US for estrogen replacement therapy). A leading gynaecologist has called HRT 'probably the most important advance in preventative medicine in the western world for half a century'. There are voices of dissent both within the medical establishment and from feminists, but rightly or wrongly, the use of HRT is now advancing almost as rapidly as that of oral contraceptives when they were first introduced.

HRT generally includes both oestrogen and progesterone. It protects against heart attack (the risk of which increases after menopause), reduces vaginal dryness and counteracts nervous symptoms. Most importantly of all it preserves bone density, thereby preventing or lessening osteoporosis and reducing the risk of bone fracture in old age. With post-menopausal bone loss affecting an estimated 80 million women in Europe, Japan and the US, demand for HRT is likely to grow. On the negative side it confers a slightly increased risk of breast cancer, but this is not severe enough to outweigh its benefits.

There are some benefits to menopause. Women may be glad to be free of the nuisance of periods and pre-menstrual syndrome. In some societies they can come out of purdah and mix freely in the world at large for the first time in their adult lives. Above all, the menopause heralds freedom from the enslavement to their own fertility and the burden of excessive childbearing that is still the lot of many women around the world.

Why has evolution allowed women to lose their reproductive ability (and therefore the opportunity to spread their genes) at an earlier age than men? It has been speculated that the menopause could have been favoured by natural selection because it allows women to stop breeding and care for their grand-children instead. This benefits the woman's own genes, albeit in diluted form, as the grand-children obtain 25% of their genes from each grandparent. It is argued that a grandparent may be of greater benefit to her family, and therefore to her own genes, than a woman who attempts to produce offspring when too elderly, when her ability to see her children through to independence would be limited.

Attractive as this theory sounds, it is likely that relatively few of our ancestors lived to reach menopause or grandmotherhood. Of the thousands of skeletal remains unearthed at ancient burial sites, none is thought to be older than 40 (though we cannot be certain). Most seem to have died in their twenties, and if this is true then evolution could not have favoured grandparents. Like so much of what we see in ageing, the menopause probably reveals no grand plan or design but is merely a mechanical failure in a body abandoned by its genes.

For men there is no equivalent to the menopause, but they do undergo hormonal and reproductive changes. Testosterone levels drop, and some authorities refer to a 'male climacteric syndrome'. However, there is no recommendation that men should receive HRT: artificially high testosterone levels would not prevent disease and might even be dangerous.

Sperm production lessens with age and their quality deteriorates, with a rise in the proportion of malformed sperm. But those sperm that are healthy are still capable of fertilising an egg, so that old men can sire offspring long after women of a similar age are infertile. This helps explain the tendency for males in many cultures to practise polygyny (multiple female partners), whereas polyandry (multiple male partners) in women may have been hindered by the fact of menopause. Nevertheless, male sexual desire usually decreases in old age, as does the ability to achieve erection. Erectal potency diminishes from almost 100% at age 20 to 25% at age 80.

Attitudes to the elderly differ between cultures. In many traditional societies old people are respected for their presumed wisdom, their knowledge of tradition, or even because they are closest to dead ancestors. Western society tends to stress the negative and regard the elderly as a burden. 'When people are well in

old age they are agreeably surprised to find that they are not unemployable, demented, asexual, sickly and in a state of increasing decrepitude', wrote the gerontologist Alex Comfort. There is some truth in both attitudes. On the one hand old age can equal incontinence, helplessness, stupidity and unhappiness. On the other hand it can mean wealth, political clout, wisdom and importance.

Because the body is such a complex and finely tuned machine, deterioration in one system has knock-on effects elsewhere. The ear deteriorates. Balance then follows suit, and coupled with impaired vision makes accidents more likely. Bones break more easily if there is a fall, and confinement to bed has more lasting effects on musculature, balance and bone strength. Longer spells in bed lead to other problems such as thrombosis or skin ulcers: in short the deterioration of the body in old age is the result of a series of vicious circles which become more problematic as time goes on. What begin as subtle changes all lead, over the years, to healthy adults becoming frail ones, to diminished capabilities in most physiological systems, and a sharply increasing vulnerability to most diseases and to death, the subject of the chapter which now follows.

Left A tribal elder from the Maprik district of New Guinea. In ancient communities these elders can command high status due to their experience and knowledge.

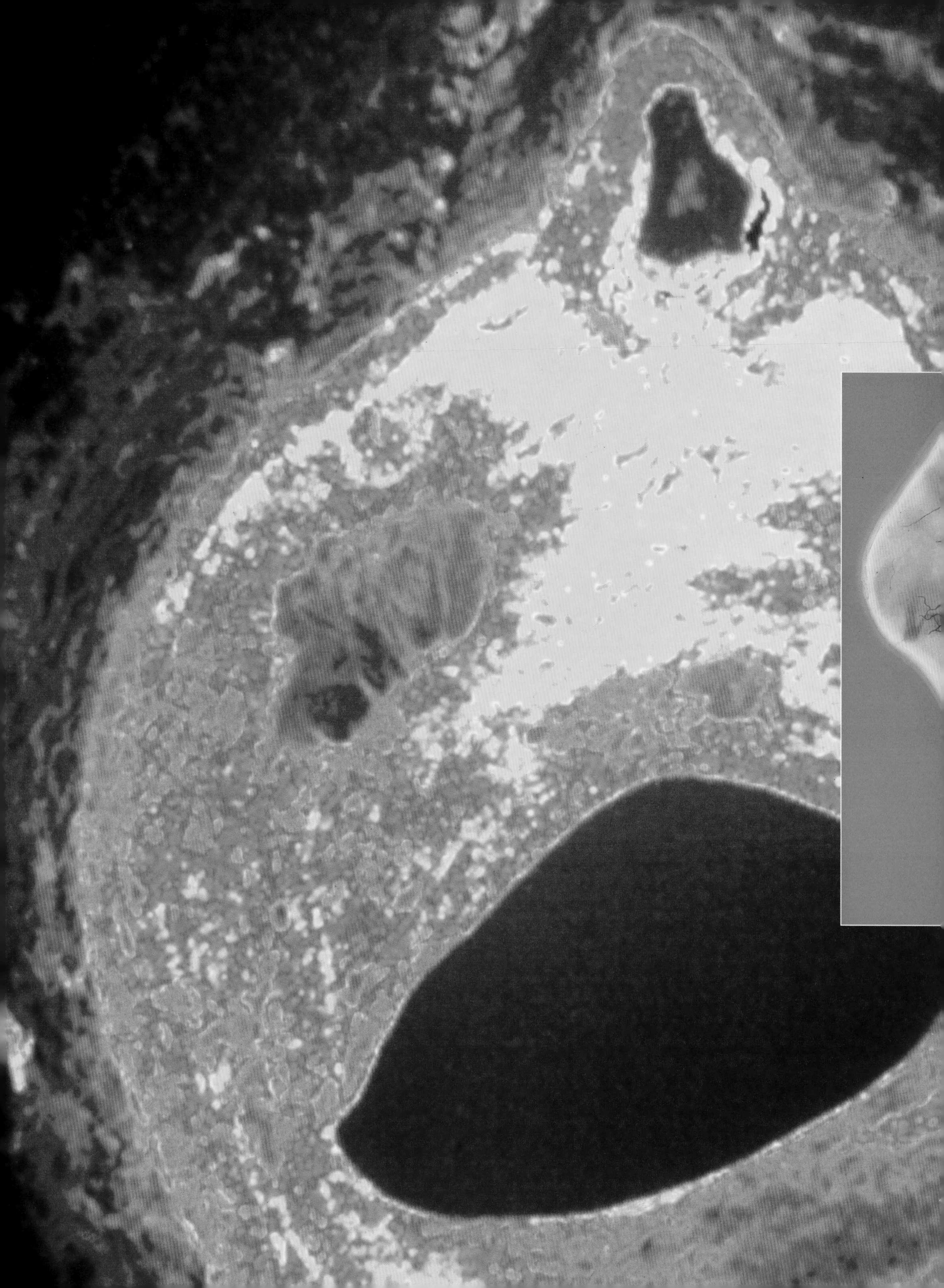

DEATH

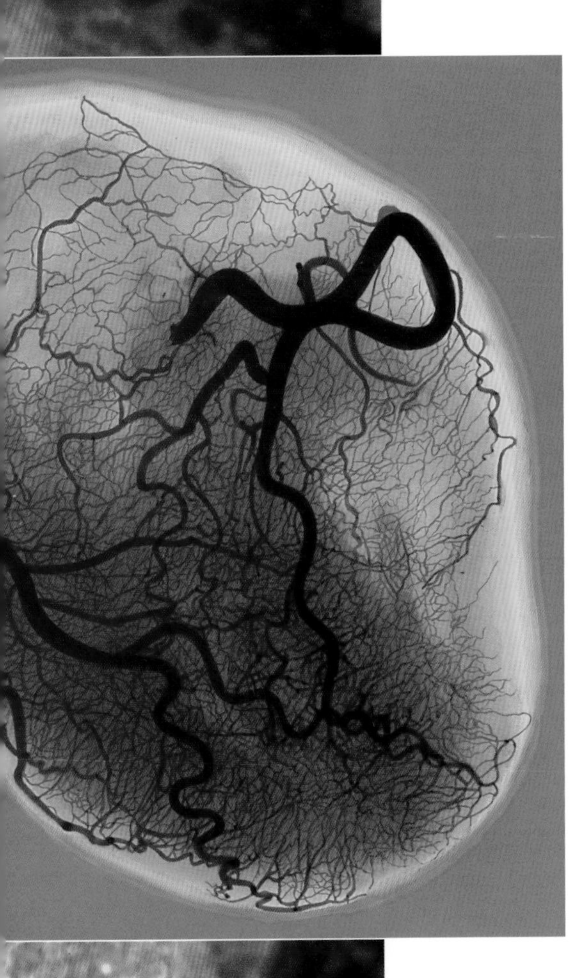

To look at an old photograph, taken before about 1895, is to know that everyone in it is now dead. To look at a modern picture, perhaps of a football crowd, is also to know that every single individual will be dead before so very long. Even to live for 100 years is only 36,500 days or 5214 weeks. With most of the football crowd already adult, and three score years and ten nearer the normal lifespan, all will be gone within another 17,000 days or so. Throughout the world 100 people die every minute. Death is not a rarity.

Yet we go about our daily lives hardly ever considering the final destination we must all reach. We think of death being much like an accident, striking here and there but not a total conqueror. Intellectually we know that, in the words of Jeanne Moreau, we suffer from 'a deadly disease which is called Life'. But we shut the thought away at the back of our minds. Much like sex when the people in the fading photograph were alive, death is now the big taboo, not much discussed and rarely witnessed.

Up until the recent past most deaths would take place at home. The family, children often included, would gather round the death bed, observing, discussing, taking note. Today in the developed world most people die in hospitals or other institutions (the figure is about 70% in Britain and 80% in the

Main picture A cross-section of an artery narrowed by atherosclerosis. Build-up of fatty plaques greatly hinders blood flow, and leaves the artery vulnerable to blockage by a blood clot which may lead to a heart attack.

Inset To detect heart disease a coloured angiogram is made. This reveals the coronary arteries and arterioles, which supply oxygen and nourishment to the heart. Any narrowing of the arteries is revealed by this technique.

HERBIE

Herbie and his wife Hannelorre retired from their native Germany and moved to live in a house he called 'paradise' in western Ireland. In 1995, after a sudden bout of severe pain, Herbie learned he had an inoperable, terminal tumour in his abdomen. Doctors gave him two to six months to live.

On hearing his prognosis Herbie had been 'a little disappointed', because he was only 60. Hannelorre was, at times, less philosophical.

Instead of the two to six months forecast, Herbie survived for 18. He lost his chubby appearance, but not much weight, because the growing tumour compensated for a shrinking frame. He resented confinement to bed when that time came, but accepted it stoically. He continued smoking, claiming he would be dead when he no longer asked for cigarettes.

On a Tuesday he announced to Hannelorre that he was dying. He was running a slight temperature, and having difficulty breathing. His lungs were filling with liquid and there were, on occasion, long pauses between breaths. The doctor thought he had another three hours to live, but Herbie was still alive when the next day dawned. All that morning his breathing became more troublesome, and he was no longer talking coherently. Occasionally his body twitched, perhaps from stabs of bedsore pain, but the individuals with him were most aware of his breathing and its growing irregularity.

At 3.30 on the Wednesday afternoon his breathing no longer wrestled in its agony, and it quietly stopped. A tremendous silence filled the room. Herbie had finally departed, from his body, from that room, and from the living who remained. It was an ending that tremendously moved all those who witnessed it.

At the hospital they removed his tumour, the thing that had caused his death. When they put it on the scales it weighed 44 lbs, or 20 kilograms. The doctors were amazed at its owner's blatant unwillingness to abandon a life which had stayed so full of zest until the certain end. By then, Herbie had passed through, to use John Webster's phrase, one of the 'ten thousand several doors' which open onto death.

US). Sometimes they are surrounded by more medical equipment than family members. We see thousands of faked deaths in films and on television but many of us will go right through life without witnessing the real event. If we do, we can be at a loss. Somerset Maugham wrote: 'To see how others have taken that final journey is the only help we have when ourselves we enter upon it.' Nowadays we do not see much of that finality, and many have lost the religious certainties that guided earlier generations. More than ever before, death is the great unknown.

Changing patterns of death

It is not only social attitudes to death that are changing. The last few generations have also seen enormous changes in the age at which people die and the causes of their deaths.

Up until the second half of the nineteenth century, infectious diseases, chiefly tuberculosis, smallpox, dysentery, typhoid and diphtheria, were the major cause of death. During the 1840s about 18% of all deaths in the UK were due to TB. Cholera was rampant in London, spread via sewage in the public water supply. Whooping cough, scarlet fever and measles claimed countless babies and children. Earlier, in the seventeenth century, bubonic plague decimated the population of Europe. Life expectancy for a child born in Western Europe in 1830 was just 40 years, and even this was an improvement on the figures prevailing two hundred years earlier. The main cause of the dramatic increase witnessed since then has been improved public health, particularly cleaner water, better sewage disposal and better nutrition. Vaccinations and effective drugs only came into their own later, when the grip of the major infectious killers was already largely broken.

Statistics on 'life expectancy' are often quoted, but can be misleading. The figure used is generally life expectancy *at birth*, the average time a newborn baby can be expected to live. Built in to this figure is infant mortality, which until recent times accounted for the majority of deaths in any one year. In the 1830s, when life expectancy was 40, few people lived to the age of 40 and then died. Those who survived the vulnerable years of infancy and childhood could look forward to a lengthy span, and those who reached 50 had almost as good a chance of seeing their 70th birthday as a 50-year-old in 1979.

Life expectancy in the developed world is now around 72 for men and 78 for women. The longest-lived are the Japanese, with male and female expectancies of 76 and 82 respectively. In the rest of the world the figure has gone from under 40 before the Second World War to about 55 in the 1990s, and is still rising. The increase has been slowest in Africa, which had a life

Cancer is the biggest killer in the Western world after heart disease and stroke. For example, in Britain, one person in three will be diagnosed with cancer at some time in their life, and one person in four will die from it. Every year in the developed world, it is diagnosed in about 1 in 250 men and 1 in 300 women.

The most common form is lung cancer, of which 90% of cases are caused by smoking. Smoking is estimated to be responsible for one-third of all cancers, due to the wide range of harmful chemicals it introduces to the body. Next most frequent is non-melanoma skin cancer, which has a 97% survival rate, followed by breast, bowel, prostate, bladder and stomach cancers.

In normal body tissues, new cells are constantly being created to replace those that have died. This is achieved through cells dividing in half to form two identical descendents. Death and renewal are kept in balance by genetic instructions governing cell behaviour. In cancer the system of regulation breaks down, and cells begin to divide uncontrollably. Sometimes the process is rapid and sometimes it may take 10, 20 or 30 years, but eventually the cancerous cells build up to form a lump known as a tumour or neoplasm (meaning 'new growth'). As the tumour grows, capillaries form around it to supply it with blood, probably in response to chemicals secreted by the tumour cells. Instead of performing a useful function, the cancer tissue greedily absorbs nutrients at the expense of the neighbouring tissues. If the tumour is on a vital organ, disruption of the organ's functions can lead to death.

The most sinister property of cancer cells is their ability to invade the surrounding tissues and spread to other parts of the body. The tiny microtubules and filaments by which cells move become disorganised, and the shape of the cell is changed. Instead of staying put it somehow pushes through its normal neighbours to extend the boundaries of the tumour. Cancer cells that reach a blood vessel or lymph duct can break free and be carried around the body, setting up new tumours where they eventually settle. This process is known as metastasis. The earlier a tumour is detected and treated, the less chance there will be that it has spread.

● *Below* A tiny lung tumour filling an alveolus, one of the hundreds of millions of air sacs that are present in each lung.

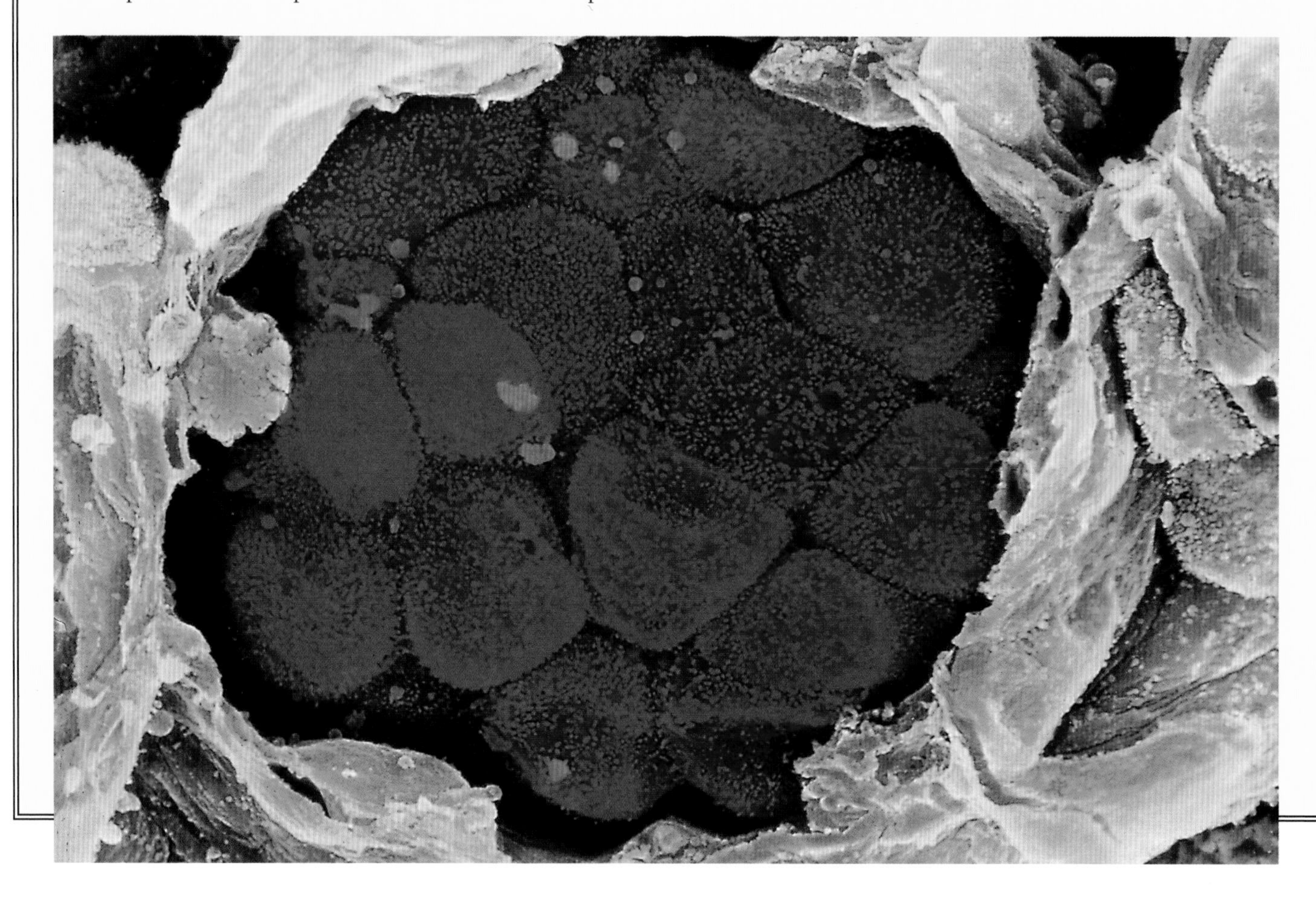

Although it can strike at any time of life, cancer is overwhelmingly a disease of old age. The majority of people who develop cancer in Britain are over 70, and figures from the US put the average age of diagnosis of the most common cancers at 67. To understand why this is, it is necessary to look at the processes that transform a normal cell into a cancerous one.

The root cause of cancer is damage to the DNA, leading to abnormal gene function. Every time a cell divides, mistakes are made in the copying of its DNA. (DNA is discussed more fully in chapter six, 'Ageing'.) Many are detected by repair mechanisms, but a few slip through and are passed on as mutations to the next generation of cells. The more mutations that build up, the more likely the cell is to become cancerous. Mutations are also caused by chemicals in the environment and in the body itself, by ultraviolet and ionising radiation, and by some viruses. Again, the longer a person lives the more damage their DNA sustains. Some individuals are born with a genetic make-up that predisposes them to certain cancers from the beginning.

Gene damage that makes a cell potentially cancerous is known as 'initiation'. For cancer to develop the initiated cell probably needs to be stimulated by an additional factor (or factors) called a 'promoter'. These events are complex and poorly understood. However, there is growing evidence that most human cancers originate from a single rogue cell.

The idea that modern societies face a cancer epidemic caused by radiation, pesticides, pollution and food additives is a myth. Research in Britain has estimated that only 2% of cancers are linked to pollution, and food additives may actually have helped reduce the incidence of stomach cancer. The most important factor in cancer risk could well be diet: Sir Richard Doll, who proved the link between cancer and smoking in the 1950s, has estimated that 35% of cancers may be diet-related. There is a lot more to learn, but current evidence suggests that fruit, vegetables, fibre and certain vitamins are protective.

The incidence of cancer is rising as the population lives longer. There are also unexplained increases in testicular cancer in young men, and in breast cancer in women. The good news is that survival rates are improving as our understanding of cancer grows.

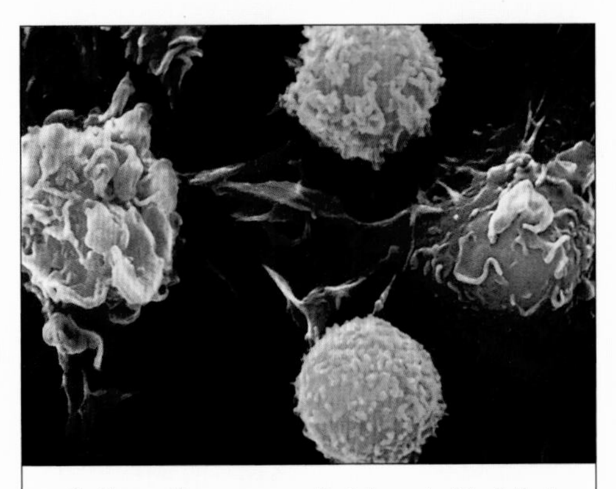

Above Two cancer cells (shown in blue) that have just divided. Cancer cells tend to have uneven surfaces, as seen here, and may use the outgrowths for movement. The pink objects are white blood cells.

Below right A cancer cell being attacked by a white blood cell (green). Certain types of white cell seek out and kill cancer cells, but offer only a partial defence against the disease.

expectancy of 47 in the mid-1970s compared with 57 for Asia and 62 for Latin America. These areas still suffer relatively high mortality from what are now termed 'tropical diseases', in many cases the very same diseases that devastated Europe in the recent past. These are not diseases of a particular climate or region, they are diseases of poverty.

In the industrialised world, records show that although life expectancy has improved dramatically, maximum *lifespan* remains virtually unchanged. As the writer of Psalm 90 knew in ancient times, 'the length of our days is seventy years, or eighty if we have the strength.' It is not 55, 60 or 90, though the 90s are being attained by increasing numbers. A few make it past 100, but they are exceptional. There is something inherent in the human body that stops us living much beyond 120. Many scientists believe that genes are involved, but as yet they have no real idea of how the system might operate. What seems certain is that life expectancy will continue to approach ever-closer to maximum lifespan for the foreseeable future.

A striking feature of the statistics, and one borne out by a visit to any residential home for the elderly, is that women live significantly longer than men. Males have a higher death rate than females at all stages from conception onwards. There are estimated to be 1.5–2 times as many male conceptions as female, and 1.3 times as many males are miscarried. There are 1.05 times as many live male births, but by the age of 30 the greater male death rate has brought the ratio back to 1:1. Women then gradually begin to outnumber men, until in the over-85 age group there are more than two for every male.

The reasons for the difference are not fully understood, but it is known to involve a complex mixture of biological and sociological factors. Biologically, men are more vulnerable to certain genetic conditions because they have only one X chromosome to a woman's two. Up until the menopause, women are less prone to heart disease because of the protective effect of oestrogen, which improves the way the body handles cholesterol. Some aspects of women's immune systems are stronger than men's, and are less affected by ageing. In evolutionary terms it has been suggested that men's bodies age more quickly than women's because male reproductive success is so dependent on competitive ability. Male physiology, it is argued, is devoted more to this competition and less to the preservation of the body. Men are also more likely than women to die from suicide, accidents and homicide, and tend to drink and smoke more. They also have more dangerous occupations, although women are exposed to the dangers of pregnancy and childbirth.

Today's crop of diseases which cut the biggest swathes in the developed world – cancer, heart disease and strokes – are only reigning so triumphantly because we are, in general, surviving other threats and living to become elderly. The traditional killers have been forced to yield. By the 1960s, combined deaths in Britain from diphtheria, measles, anthrax, dysentery, polio, smallpox, whooping cough, tuberculosis, cholera and plague – all horrific names from the past – totalled less than half the number from road traffic accidents. When one hazard is conquered another steps into its place. The car now kills as many worldwide as measles or malaria.

The figures in the panel on page 192 were compiled for a Global Burden of Disease study by the World Bank and the World Health Organisation. They relate to the year 1990. Suicides outnumbered death from AIDS in that year (although AIDS figures have risen since then), and drowning killed more than war. Four of the top ten (heart disease, stroke, obstructive lung disease and respiratory cancers) could be partly prevented, according to the study, by a reduction in tobacco smoking, taking more exercise, cleaning up industrial pollution and improving diets. Measles could be conquered by immunisation.

By the year 2000 it is expected that obstructive lung disease will jump from number 6 to number 3, respiratory cancers from number 10 to number 5, road traffic accidents from number 9 to number 6, and self-inflicted injury from number 12 to number 10. Obstructive lung disease, which leads to heart failure, will rise because it is essentially cigarette-induced and smoking in the developing world is on the increase. Number 3, lower respiratory infections, may be displaced as living conditions in developing countries improve and more people gain access to antibiotics. The same improvements will push infectious diseases and malnutrition down the league and boost the death rate from heart disease and stroke, the big killers of old age and affluence. Suicide is expected to rise.

MAJOR CAUSES OF DEATH WORLDWIDE, 1990

RANK	CAUSE OF DEATH	NUMBER OF DEATHS
1	Ischaemic heart disease	6.26 million
2	Cerebrovascular disease (stroke)	4.38 million
3	Lower respiratory (lung) infections	4.30 million
4	Diarrhoeal diseases	2.95 million
5	Perinatal disorders (death shortly before or a few weeks after birth)	2.44 million
6	Chronic obstructive lung disease	2.21 million
7	Tuberculosis	1.96 million
8	Measles	1.06 million
9	Road traffic accidents	999,000
10	Trachea, bronchus and lung cancers	945,000
11	Malaria	856,000
12	Self-inflicted injuries	786,000
13	Cirrhosis of the liver	779,000
14	Stomach cancer	752,000
15	Congenital anomalies (birth defects)	589,000
16	Diabetes mellitus	571,000
17	Violence	563,000
18	Tetanus	542,000
19	Nephritis and nephrosis (diseases of the kidney)	536,000
20	Drowning	504,000
21	War injuries	502,000
22	Liver cancer	501,000
23	Inflammatory heart diseases	495,000
24	Colon and rectum cancers	472,000
25	Protein–energy malnutrition	372,000
26	Oesophageal cancer	358,000
27	Pertussis (whooping cough)	347,000
28	Rheumatic heart disease	340,000
29	Breast cancer	322,000
30	AIDS	312,000

If asked how we would like to die, many of us would reply 'from old age'. Instead of suffering disease prior to death we would like to remain reasonably healthy until one day our body stops working, preferably in our sleep. But does anyone really die of old age, or is there always an underlying disease?

The World Health Organisation's International Classification of Diseases gives little opportunity to classify old age as a cause of death. The closest it gets is a category for 'senility without dementia', but in 1986 this was assigned to only 0.3% of deaths in the over-85s. However, a review of 200 autopsies on Americans who died at 85 or older was unable to find a cause of death in 26% of cases. The researchers concluded that these people died from diseases which in younger people are not life-threatening. Others believe that a large proportion of deaths in the elderly cannot be accounted for by disease at all, but what might trigger death from 'old age' remains a mystery.

Death as a part of life

'In the midst of life we are in death'. This pronouncement from the Christian burial rite is also true in a biological sense. At the cellular level, death is integral to life itself (see pages 194 and 195). Every day 2% of our blood cells die and are replaced by fresh ones. The cells lining the gut die off and are renewed every third day, and we are constantly shedding dead skin cells (which account for about 75% of household dust). Bone cells are in perpetual turnover, so that our whole skeleton is replaced every seven years.

Cells also die without being replaced. By the second half of our lives we are losing thousands of brain cells (neurones) every day, the brain losing around 2% of its weight every decade after 50. As many as half the neurones in some crucial areas of the brain disappear during a lifetime. Cells in the cardiac muscle also die off, and the sino-atrial node, the heart's pacemaker, can lose 90% of its cells by age 75. Each kidney loses 20% of its weight between 40 and 80, by which time half of its filtering system will have stopped functioning.

PROGRAMMED CELL DEATH

Nobody knows the exact numbers, but scientists are agreed that many millions of cells die in our bodies each day. Some die from injury, but for most, the cause of death is suicide.

Programmed cell death, known as apoptosis from the Greek word for 'falling off', has become a major research topic in biology. We now know that cells' ability to self-destruct to order is vital to the smooth functioning of the body. Failure of the system is implicated in causing cancer and rheumatoid arthritis, while too much apoptosis is thought to contribute to AIDS, Alzheimer's and Parkinson's disease.

When a cell dies from injury, its capacity to control its fluid balance is lost and it swells up and ruptures. This process is known as necrosis, and is usually accompanied by inflammation of the surrounding tissue. Programmed cell death is quite different. Instead of swelling, the cell shrinks. Its surface begins to look as though it is boiling. Inside, the nucleus undergoes changes not seen in necrosis. The dying cell eventually falls into a number of pieces and is mopped up by special scavenger cells, bypassing the inflammatory response so that cells around it are not damaged. Sometimes the dead cell remains intact and is fossilised by the invasion of particular proteins. For example, the lens of the eye is made up of dead cells that are flooded with the protein crystallin.

Apoptosis was first demonstrated in the foetus. In order to develop properly, embryos of all species must kill off certain groups of cells. Their progress is shaped by a constant process of over-creation and selective elimination, honing crude cell clusters into intricate organs and structures. This is particularly important in the brain, where neurones that fail to form connections with others are pruned out by the million. The process also sculpts the human hand, paring back the mitten-like web of skin that joins the fingers when they first form. What triggers this no one knows. Other examples of apoptosis include the turnover of cells in the skin and gut lining, the destruction of potentially dangerous white blood cells in the thymus, and the monthly shedding of the cells of the uterine wall.

Cell suicide is carried out using deadly, protein-destroying enzymes that the cell stores in deactivated form until the trigger to use them arrives. One such trigger is the failure to receive chemical signals from other cells. It is well known to scientists that a single cell placed on its own in a culture medium will die, whereas a group of cells under the same conditions will survive. Without constant signals from their neighbours urging them to stay alive, cells will simply do away with themselves. Sometimes these signals can be overridden by others, either from within the cell or outside it, with the same end result – suicide. In a healthy body, any cell that becomes dangerous (perhaps through genetic mutation or infection with a virus), or simply superfluous, will do the decent thing and sacrifice itself for the greater good.

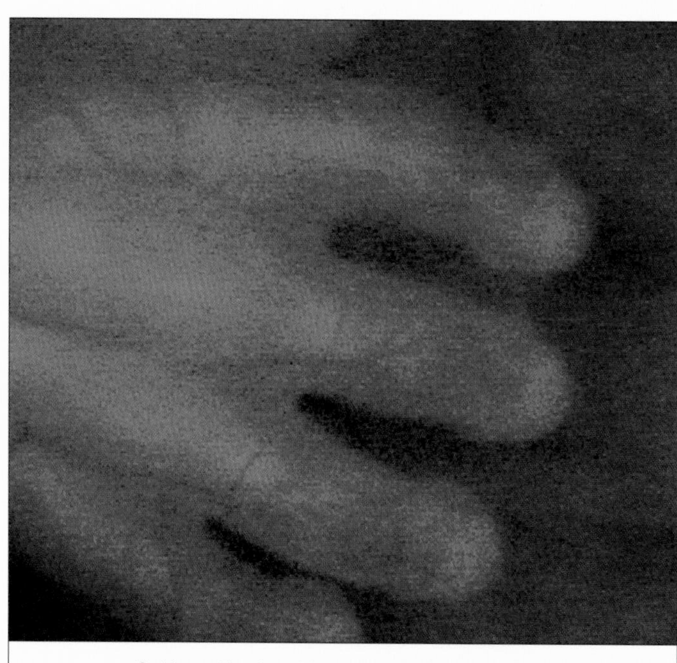

● *Above* The foetal hand sculpted by apoptosis.

● *Right* A white blood cell undergoing apoptosis (centre). The cell has shrivelled into grape-like blisters and will soon fall apart. The cells surrounding it are normal.

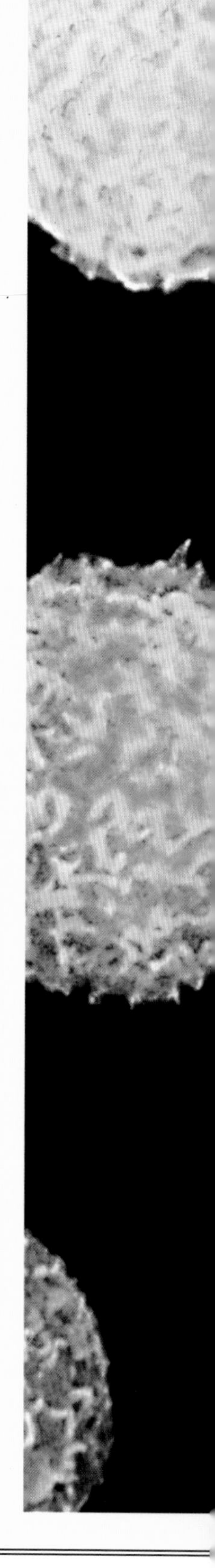

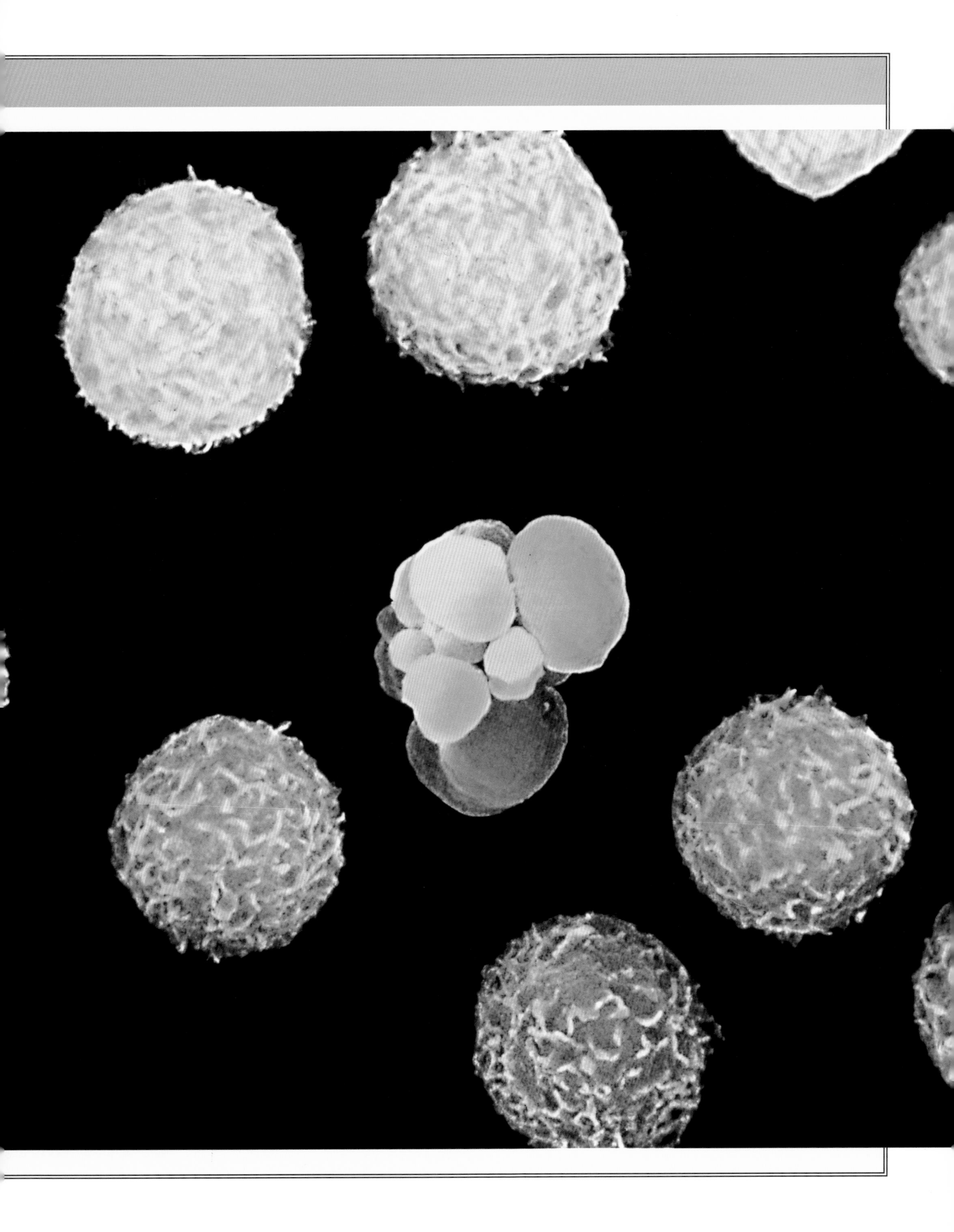

Therefore to some extent death is a process, with the final death of the whole as its culmination.

Nature has none of our qualms about death. The lifespan of most wild creatures is cut brutally short by predation, starvation or disease, or simply because the individual is too weak, perhaps through genetic abnormality, to survive the challenges it faces. The weakest die before breeding, and their genes die with them. Those poorly adapted to their environment are likely to die sooner and breed less than well-adapted individuals. Death is thus a tool of evolution.

It is more difficult to imagine this tool at work in the human race today, especially in the developed world. Plentiful food, lack of predators and modern medical techniques mean that the great majority of those who are born will survive to adulthood and reproduce, including many who would have died just a century ago. But those who are born at all have already come through a stringent process of weeding out. Of all the potential human beings created every time a sperm fertilises an egg, perhaps 50% or even more never survive to create fertilised eggs of their own.

Many fertilised eggs, perhaps most, fail to implant in the uterus and are swept out with the menstrual flow. Studies using sensitive hormone monitoring have put the loss of conceptions up to the 12th week of pregnancy at between 31% and 62%, with the majority occurring before the woman realises she is pregnant. To a large extent this seemingly enormous wastage of potential life is a pruning out of unviable genetic combinations, and a large proportion of deaths in infancy also have a genetic cause.

Oxygen – the vital ingredient

The single crucial factor in all human (and animal) death, whatever the malfunction that leads to it, is lack of oxygen. Whether the lungs fail to obtain it, the heart stops pumping the blood that carries it, or the brain can no longer coordinate its delivery, if our tissues stop receiving oxygen we are dead. As circulation, respiration and neuronal control are intimately linked, failure in one quickly sets up a vicious circle that drags down the others.

The brain, the most critical organ of all, is also the most demanding of oxygen. It uses 20% of the body's oxygen intake but accounts for only 2% of body weight. In adults it will die if deprived of oxygen for four minutes. (Strangely it is also the most sensitive of all tissues to toxicity from an excess of oxygen.) Other tissues can survive a little longer, but all rely absolutely on oxygen for life. Without oxygen the body cannot burn food to release energy, and without energy the body winds down and stops – like a clockwork toy.

We generally associate the word 'respiration' with breathing, the physical inhalation and exhalation of air from the lungs. But breathing is only half the process. The oxygen that enters the blood at the lung is delivered via tiny capillaries to individual cells, where it is used in the chemical reactions that are termed 'internal' or 'cellular' respiration.

Once it has diffused into the cell the oxygen reaches the mitochondria, tiny yet complex structures that are just one example of the miniature organs found in this microscopic world. A mitochondrion is about three one-thousandths of a millimetre long, but the electron microscope has allowed it to be viewed and photographed in stunning detail (see pages 198 and 199).

Inside the mitochondria the oxygen molecules take part in an intricate series of chemical reactions with 'fuel' molecules derived from food. Energy is released from the food molecules and stored in the form of chemical bonds in a molecule called ATP. ATP molecules move around the cell and give up the energy when it is needed, perhaps to contract a muscle, send a nerve signal or manufacture a hormone. This process is the fundamental driving force behind everything the body does, and can be summarised as an equation: food + oxygen = carbon dioxide + water + energy. The carbon dioxide and water are waste products. Carbon dioxide is exhaled via the lungs as is some of the water; any excess water will be excreted by the kidneys.

The brain's voracious appetite for oxygen can be quelled temporarily by lowering its temperature. Normally such an act would be fatal, but on the

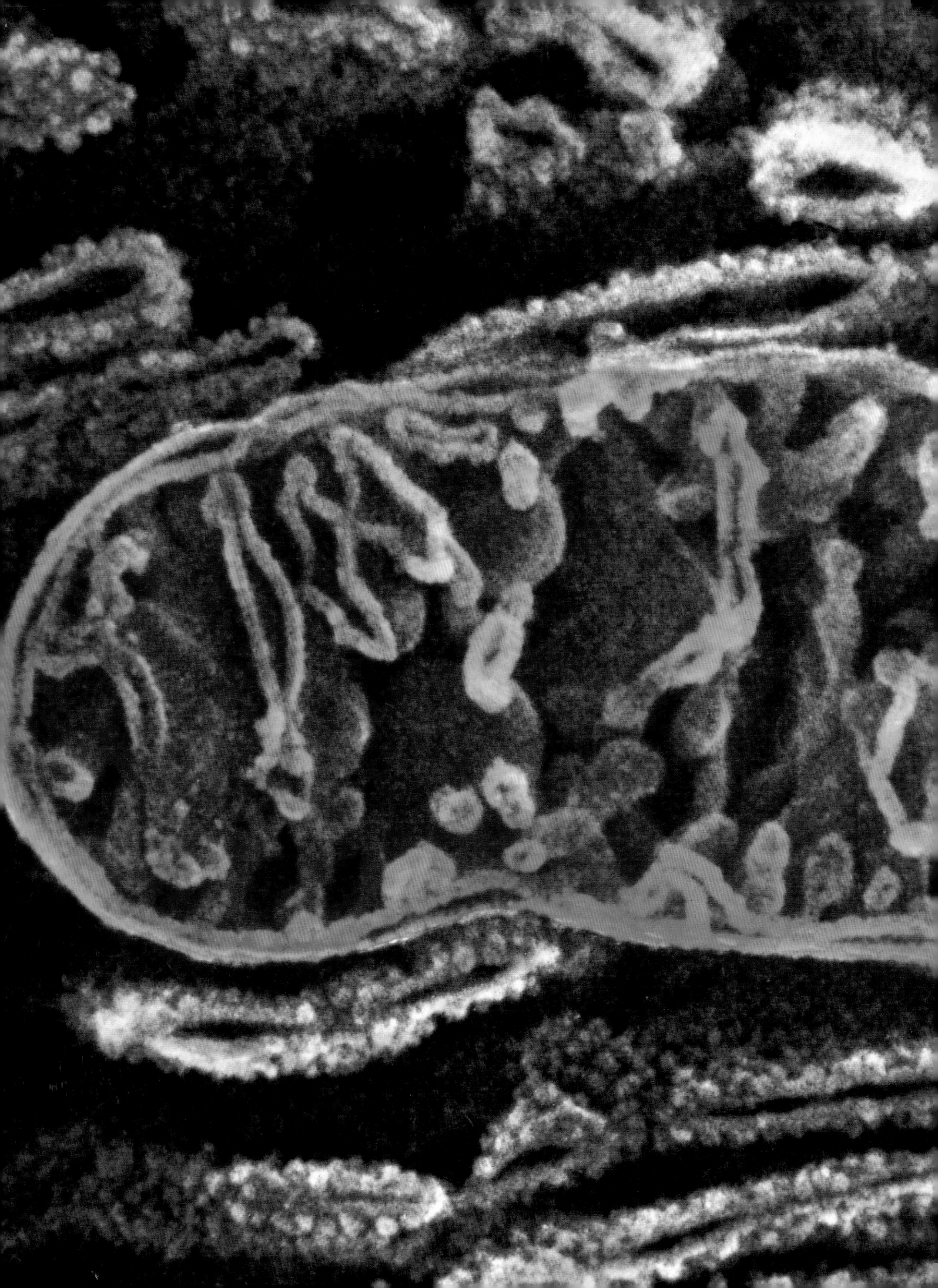

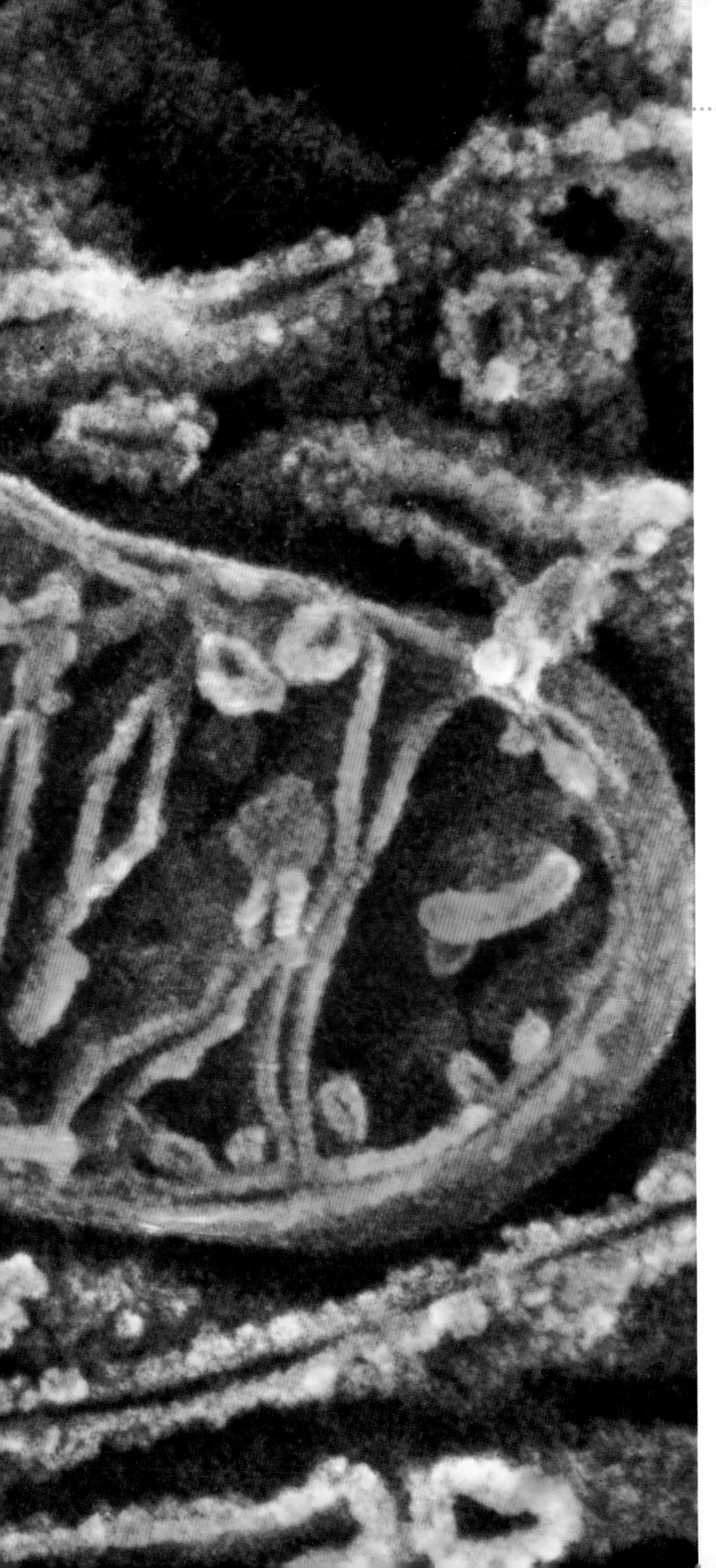

Left Section through a mitochondrion. These structures occur within every cell, and are the sites of cellular respiration. They burn sugars and fats to release the energy needed to drive all the body's functions and activities.

operating table it has proved advantageous. In the 1960s a Japanese neurosurgeon cooled brains to 6°C (43°F) from their normal 37°C. After chilling them with ice to 32°C he drained their blood and replaced it with a cold substitute. He was then able to operate on a cold, white, bloodless brain, which he rewarmed afterwards with a transfusion of warm blood. The brain's 'revitalisation' as he termed it took about one hour. The six individuals treated in this way suffered no reported harmful effects.

Cryosurgery (cryo- meaning extreme cold, from the Greek word for frost) is now used in the removal of aneurysms, swellings in the blood vessels of the brain that can be fatal if they burst. Some large aneurysms cannot be operated on under normal conditions because the risk of bursting is too high. The swelling must be drained of blood, which means draining about half the blood from the patient's body. To prevent tissue damage from lack of oxygen, the whole body is cooled to about 18°C by packing it with ice. At this temperature, brain activity is greatly reduced, all the vital signs disappear and the tissues require very little oxygen to stay alive. The patient is in a state of suspended animation where the brain can survive up to an hour without oxygen, as against four minutes at normal body temperature.

Diagnosing death

The body under cryosurgery has no blood flow, no heartbeat, and no reactions. Anyone seeing such a body under other circumstances would consider it dead, and by many definitions it is, but for the crucial fact that it can be brought back to life again. Modern medicine also creates bodies that, although the heart beats and air flows through the lungs, are pronounced dead. Death is the ultimate finality yet it has no simple definition or diagnosis.

Until quite recently someone was pronounced dead once they seemed to have stopped breathing. Breath has been regarded as the essence of life since ancient times: the word for soul in Sanskrit is the same as that for breath, and the Greeks called the vital spirit of life 'pneuma', the word for air. When the stethoscope was invented the emphasis switched to the stopping of the heartbeat, but the diagnosis of death was still not foolproof. People were sufficiently worried about being buried alive for various strange inventions to surface to prevent such a mistake. Some coffins were fitted with bells for the undead to make their presence heard. A Dr Jost won a prize from the Academy of France in the nineteenth century for inventing nipple-pinching forceps to prove an absence of sensation. An English invention involved painting the words 'I am dead' in silver nitrate on a glass pane. The words would remain invisible until the decomposing corpse gave off enough hydrogen sulphide gas to change the silver nitrate into visible silver sulphide. Funerals were often delayed until putrefaction became evident. The horror writer Edgar Allan Poe apparently had much to do with encouraging this fear of wrongful burial.

Conditions such as hypothermia, some drug overdoses, metabolic disturbances and coma can mimic death, and even today stories surface from time to time of mortuary attendants noticing a twitch from a supposed cadaver. The majority of deaths though are plain to see and involve no doubt. Hospice workers often speak of departure, of sensing that something goes. Ophthalmologists, peering into an eye, have seen blood corpuscles rolling to a stop. Pulse and breathing stop, and other signs of death soon become apparent. The facial muscles sag. The eyes stare, taking on a new look and losing their sheen. The pupils dilate to an enormous size, and the round plumpness of each eye becomes a flatness. Skin whitened by pressure does not readily regain colour, and there is no bleeding if the skin is cut or blistering if it is burnt. A purple stain appears where blood settles in the lower parts of the body.

In a straightforward case such as the death of a terminally ill patient in a hospice, doctors will confirm death simply by a prolonged search for a pulse followed by a second check some time later. Where death is unexpected there is the possibility that the person may appear dead but be capable of resuscitation. In cases like drowning or sudden collapse it is vital to attempt resuscitation first and ask questions later, as precious time could be lost searching for evidence of death. Otherwise, the doctor will check thoroughly for pulse and respiration at various sites, examine the eye and check it for reflexes, and look for the other signs mentioned above.

In modern times there is a third category of death, where diagnosis is controversial both medically and ethically. This issue of 'brain death' is explored below.

Brain death

In a minority of deaths, efforts to save the patient involve artificial life support, including the use of a ventilator to keep the body supplied with oxygen. Modern technology means that all the vital functions can be kept going in a body whose brain is irreversibly damaged. In a bizarre reversal of the natural death process, the death of the brain takes place before, rather than after, the collapse of the other systems.

When it is concluded that a person on life support will never be capable of recovery, the decision can be made to withdraw or switch off the equipment. This act will stop the flow of oxygen and cause the still-living tissues to die, destroying once and for all any possibility of life. Before going ahead doctors must establish brain death, defined as the death of the brain stem. The stem is the most primitive part of the brain, and many of its functions are similar in all vertebrate animals. It controls breathing and heartbeat, regulates our sleeping and waking patterns and performs many of the myriad small adjustments that maintain our internal environment. Irreversible damage to the brain stem causes an irreversible loss of the ability to breathe naturally and of the capacity for consciousness. It is the point of no return.

Determining brain stem death is not straightforward, because there may still be activity in other brain regions. Various sets of criteria have been drawn up, the best-known of which is the Harvard list of 1968.

Under the Harvard criteria the patient should be:

- completely unresponsive to painful stimuli
- without movement for one hour
- without spontaneous breathing for three minutes with the respirator off
- without brain stem reflexes, such as responses to light shone in the eye, touching of the cornea and passage of a tube down the throat.

A flat electroencephalogram (EEG) reading is a useful confirmation but is not in itself diagnostic. All the tests must be carried out again 24 hours later, and the results must be unchanged.

Yet another area of uncertainty is the so-called persistent vegetative state (PVS), an unrousable coma in which the brain stem is still functioning and the person can therefore breathe unaided. There may be some reflex movements, and the eyes may open and move about, but there is no activity in the higher parts of the brain associated with consciousness. If the patient is fed (by tube), the body can carry on in this state for years. There are currently some 14,000 people in PVS in the USA alone.

Is the person in PVS alive in any meaningful way, or is the act of feeding them a grotesque intervention that is keeping them from a dignified death? Opinion is split. The person is not dead by any current medical definition, but neither is he or she capable of recovery. In the UK relatives of some PVS patients have successfully argued in court that feeding is a medical intervention and have succeeded, with the cooperation of medical staff, in having it legally withdrawn. Others believe this is unethical, and cite cases where patients thought to be in PVS have turned out to have awareness.

In all these situations our traditional understanding of what is meant by being alive becomes inadequate, and we enter new and unresolved territory.

Disease of the heart and circulation is top of the league of killers, accounting for one-fifth of all deaths worldwide and nearly half of those in the Western world. But the human heart is far from weak. At 70 beats per minute the heart is driving 8000 litres of blood around the body every 24 hours, each corpuscle coming back to the heart in just over a minute. Furthermore, there is no let-up for the heart. It has to maintain itself and continue its life-long exercise without faltering, let alone resting. To hesitate is to put other organs at immediate risk, notably the brain. Each human heart will contract 2.5 billion times in an average lifetime.

There are two main forms of heart and circulatory disease. Ischaemic or coronary heart disease is the greatest killer, with cerebrovascular disease or stroke a close second. Ischaemic heart disease (the word means 'insufficient blood supply') affects the supply of blood to the heart muscle itself. The heart is not nourished by the blood passing through its chambers but by blood in a network of coronary arteries wrapped around the outside.

The coronary arteries are susceptible to the build-up of fatty deposits, a process that can start in childhood but becomes significant in middle and old age. Age also causes the artery walls to become thicker and less elastic – a process known as arteriosclerosis. The fatty deposits accumulate cells and connective tissue and consolidate into hard plaques called atheroma, which partially block the artery (page 184). The combination of these two conditions is termed atherosclerosis, and is the precursor to heart attack and angina.

Plaque build-up is accelerated by a diet rich in fats, especially the so-called 'saturated' fats of animal origin. Saturated fat is converted within the body to cholesterol, which is a major component of plaques. The way cholesterol is handled varies between individuals, some are far better at converting it to harmless forms than others. Susceptibility to atherosclerosis has a strong association with genetic factors, and the tendency to heart disease runs in families.

Once the arteries around the heart narrow there is a danger that they will become blocked. A piece of plaque is dislodged, further obstructing the flow, and a blood clot forms around it. Such a blockage is known medically as a myocardial infarction: in common terms, a heart attack. Suddenly blood can no longer reach a portion of the heart muscle, depriving it of vital oxygen. Within minutes, the affected portion dies. The entire heart, if unable to compensate, may then stop or begin to beat imperfectly. Unified beats may be replaced by an uncoordinated quivering known as fibrillation. The correct rhythm can sometimes be restored by electric shocks to the chest.

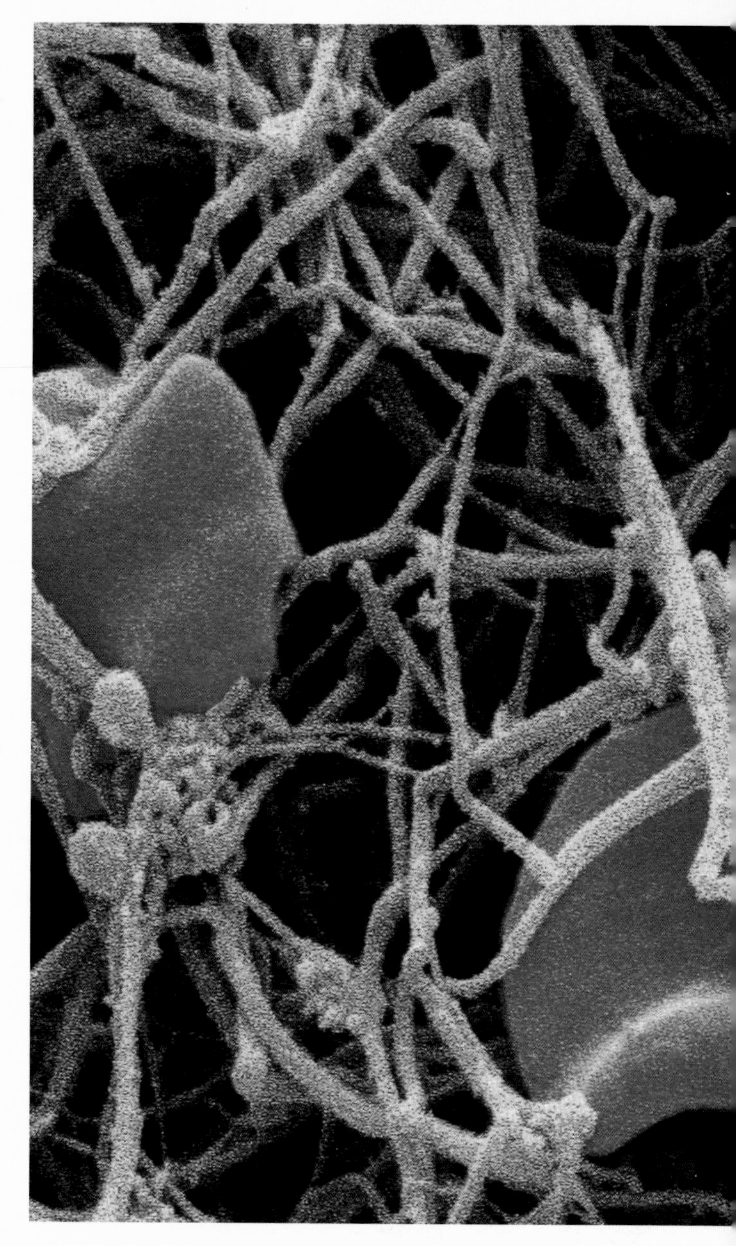

● *Above* Close-up of a blood clot with red blood cells caught in the fibrous mesh.

● *Right* Scan of a brain affected by stroke. The blue area is dead tissue, killed by oxygen deprivation when its blood supply was reduced by a clot.

The dead portion of heart muscle eventually forms scar tissue, reducing the efficiency of the heart and making the surviving muscle work harder. Sometimes the downward process is so gradual, with more blockages, more scarring and less functioning cardiac muscle, that there is not so much a heart attack as a cumulative loss of capability. Narrowed arteries may cause angina – intermittent bouts of pain from inadequate oxygen supply.

Although about 40% of 'coronaries' are of the more gradual variety, the image associated with heart attack is of sudden, dramatic collapse. The starving muscle exerts severe pain, typically with a feeling of crushing or constriction, that may radiate to the neck and arms. There is pallor, clamminess and sweating. If the victim survives to reach medical help, there is a good chance that his life will be saved, and modern treatments such as clot-busting drugs have helped reduce the chances of a second attack.

The second major killer is cerebrovascular disease. Its common name, stroke, implies a sudden disabling attack. However, like heart attacks, strokes can be small, with no immediate symptoms, or massive, when they usually induce coma. A stroke occurs when a portion of the brain is deprived of oxygen by a blockage in a blood vessel. This results in brain damage ranging in scale from minor to fatal. Someone who survives a major stroke often has sensory loss and muscle weakness on the opposite side of the body to the damaged region of the brain. They may lose the ability to speak, but not necessarily the ability to comprehend. There may be partial paralysis, and balance or breathing may be affected. A series of small strokes can cause accumulated brain damage that eventually leads to dementia of a similar type to Alzheimer's disease.

The final moments

When death is very sudden the body may be thrown in a matter of seconds from excellent health to a state where life can no longer be sustained. However, most deaths come more slowly and involve a gradual failure of bodily systems. Close to death the breathing becomes irregular, with very quick, deep breaths being followed by very slow ones. This is called Cheyne-Stokes respiration and is caused by the receptors near the heart and in the brain stem responding sluggishly to changes in oxygen and carbon dioxide in the blood.

Breaths may be noisy because the person is too weak to cough and clear the mucus from the throat and airways. High blood pressure can force fluid from the blood through the capillary walls of the lungs, which become congested. The last few breaths at the moment of death often feature a rasping sound known as the death rattle, caused by spasms in the muscles of the vocal cords. Occasionally a violent tightening of the laryngeal muscles will produce a sudden barking sound.

Immediately before death there is often a short period termed the agonal phase (from the Greek agon-, meaning struggle). The muscles may jerk in spasms caused by blood acidity, and sometimes there will be a brief convulsion or heaving of the chest or shoulders. Breaths may come in enormous gasps. Death agonies are distressing for observers, but the dying person is usually too ill to be aware of them.

At death the pupils become dilated as the muscles that control the iris relax for the last time. Occasionally the eye sheds a terminal tear, known as the *lacrima mortis*.

The body

When the heart stops pumping blood around the body, the supply of oxygen to the tissues ceases and they begin to die. Brain cells are irreversibly damaged after at most four minutes without oxygen, but cells in the muscles and connective tissues can live on for up to 30 minutes after the heart stops, using anaerobic (oxygen-free) respiration. Anaerobic respiration is also called upon in life during strenuous exercise, when muscles are worked so hard that the oxygen supply from the lungs cannot keep up with them. Low oxygen levels cause a secondary set of chemical reactions to start up to keep the cell supplied with energy, but these generate acidic by-products which quickly limit the capacity for further work. In life the acid is made harmless once oxygen levels rise: we pant after exercise to get enough oxygen to dispose of it. In death the by-products are not removed, and their build-up in the cells causes the muscle fibres to contract and become locked, the condition known as rigor mortis. The stiffness starts about six hours after death, beginning in the eyelids and jaws and spreading to cover the whole body by hour 12. As the cells deteriorate further the chemistry changes again, and the muscles relax for the final time about 24 hours after death.

The dead body cools steadily and is soon noticeably cold to the touch. A clothed body will reach ambient temperature in about 24 hours, and a naked one will cool 50% faster.

Decomposition becomes evident after 48 hours or so in a mild climate, or more quickly if the weather is warm. It first appears as a green colouration on the abdomen, which darkens to purple and then black. The early stages of decay are caused by the digestive enzymes in the gut, the body literally digesting itself from the inside. Gut bacteria soon take over and the body putrefies. When a body is embalmed for preservation or mummification the intestines are removed to slow down decay.

Bacteria need moisture, so if the air is hot and dry enough (above 38°C or 100°F) the body will not putrefy but will simply dry out. It can also, of course, be preserved by freezing, either at the morgue or naturally as with Otzi, the prehistoric 'ice man' recently found in the Alps. Even the contents of Otzi's stomach were preserved, enabling scientists to discover what he ate for his last meal. Acidity can also prevent bacterial decay, as happened with the prehistoric man christened Peat Bog after the marsh where his remains were found.

Under normal conditions, coffins and urns permitting, the constituents of the decayed or burnt body are recycled, re-entering the food chain as nutrients for plants or soil animals. Earth to earth, ashes to ashes and dust to dust, as Christians are reminded at their burial rites. Our bodies are made up of atoms that have existed since the beginning of the universe: it is a fundamental law of physics that matter can neither be created nor destroyed. The same atoms that once formed our star and then our planet are taken up in food, air and water to briefly form part of our bodies, and when we die they are handed back to be taken up elsewhere.

Glimpses beyond death

No one can know what it is like to die, but a few individuals claim to have caught a premature glimpse. 'Near Death Experiences', a term coined in 1975, are one result of the modern ability to snatch patients back from the brink of what, until recently, would have been death. However, the phenomenon has an ancient history. Plato told of Er, a soldier thought to have died who then lived again. Today, some 30–40% of those 'brought back to life' by medical intervention have reported NDEs.

Their commonest experience is a sense of peace while travelling down a tunnel towards a bright light. This joy is coupled with a choice: either return to ordinary life or stay in the lovely world to which they are heading. Some claim to have met and spoken with dead relatives. Another version is the life review, the famous flashing of streams of memories through the mind. This is most common in traumatic situations like near drowning. Still others describe an 'out of body experience', a feeling that they have left the body and are looking down on events from a detached perspective.

There have been too many such recollections with too many similarities for them to be dismissed out of hand. There does appear to be a consistent set of events perceived by some people who come close to death; the debate is over what causes them. Many who have had NDEs believe that they prove the existence of some kind of life after death, of a continuing consciousness that is independent of the body. There is often a profound change in survivors' values and beliefs, and sometimes a religious conversion. The scientific explanation is that NDEs are the brain's response to the physiological crisis at hand.

Fighter pilots have reported out of body experiences when suffering from oxygen deprivation at high G forces. Very sudden oxygen starvation is known to be associated with positive emotions, sensations of floating and inability to communicate. Under normal conditions an inhibitory system in the brain prevents too much neuronal activity. With insufficient oxygen this system may start to fail, permitting cells to fire uncontrollably. The centre of the visual field has more neurones than the periphery, so it makes sense that there should be a central bright light fading towards the edges, the standard tunnel experience of NDEs. Loss of peripheral vision – the tunnel again – is another well-known effect of anoxia. Injury causes the brain to be flooded with endorphins and other chemical messengers which mimic the effects of opiates (the chemical family that includes morphine and heroin). In large quantities these can produce hallucinations and an overwhelming sense of well-being, which could explain both NDEs and the peaceful expression often worn by victims of traumatic death.

Mankind has many cultural and religious beliefs about an immortal 'soul', which is either reincarnated or journeys to an afterlife. The only certain form of immortality is the genetic inheritance that we received from our ancestors and may or may not pass on to descendants.

Bibliography

ASIMOV, ISAAC, *The Human Body.* Signet Science Library, New York, US, 1964.

AUSTIN, C.R., & SHORT, R.V. (eds), *Embryonic and Fetal Development.* Cambridge University Press, Cambridge, 1972.

BAKER, PAUL T., & WEINER, J.S. (eds), *The Biology of Human Adaptability.* Clarendon Press, Oxford, 1967.

BANCROFT, JOHN, & MACHOVER REINISCH, JUNE (eds), *Adolescence and Puberty.* Oxford University Press, Oxford, 1990.

BENNETT, GERALD C.J., & EBRAHIM, SHAH, *Health Care in Old Age.* Edward Arnold, London, 1995.

BEST, C.H., & TAYLOR, N.B., *The Living Body.* Chapman & Hall Ltd, London, 1964.

BROOKBANK, JOHN W., *The Biology of Ageing.* Harper & Row, London, 1990.

BRYAN, ELIZABETH, *Twins, Triplets and More.* Penguin Books Ltd, London, 1992.

BURNHAM, S., WALKER, BOYD, WILLIAM C., ASIMOV, ISAAC, *Biochemistry and Human Metabolism.* Williams and Wilkins, Baltimore, US, 1957.

CHAMBERLAIN, GEOFFREY, DEWHURST, JOHN, HARVEY, DAVID, *Obstetrics.* Gower Medical Publishing, London, 1991.

COHEN, JACK, *Reproduction.* Butterworth Heinemann UK, Oxford, 1977.

COMFORT, ALEX, *Ageing; The Biology of Senescence.* Routledge & Kegan Paul, London, 1964.

Concise Medical Dictionary. Oxford University Press, Oxford, 1996.

DAWKINS, RICHARD, *The Selfish Gene.* Oxford University Press, Oxford, 1989.

DIAMOND, JARED, *The Rise and Fall of the Third Chimpanzee.* Vintage, London, 1991.

ENGLAND, MARJORIE A., *Life Before Birth.* Mosby Wolfe, London, 1996.

GESELL, ARNOLD, *The First Five Years of Life.* Methuen Academic, Routledge, 1963.

GREEN, J.H., *An Introduction to Human Physiology.* Oxford University Press, Oxford, 1963.

GUTHRIE, DOUGLAS, *A History of Medicine.* Thomas Nelson & Sons Ltd, Surrey, 1960.

HALLIDAY, TIM, *Sexual Strategy.* Oxford University Press, Oxford, 1980.

HILTON, TESSA, WITH MESSENGER, MAIRE, *Baby and Child Care.* The Bodley Head, London, 1993.

HMSO BOOKS, *On the State of the Public Health.* Norwich, 1996.

ILLINGWORTH, RONALD S., *The Normal Child.* Churchill Livingstone, Edinburgh, 1993.

JONES, STEVE, *The Language of the Genes.* Flamingo, London, 1994.

KEELE, CYRIL A., & NEIL, ERIC, *Samson Wright's Applied Physiology.* Oxford University Press, Oxford, 1964.

LARSEN, WILLIAM J., *Human Embryology.* Churchill Livingstone, Edinburgh, 1993.

LEAKEY, RICHARD E., & LEWIN, ROGER, *Origins.* Macdonald and Jane's, London, 1977.

LEVITAN, MAX, & MONTAGU, ASHLEY, *Textbook of Human Genetics.* Oxford University Press, Oxford, 1971.

LOCKHART, R.D., HAMILTON, G.F. & FYFE, F.W., *Anatomy of the Human Body.* Faber & Faber Ltd, London, 1959.

MACDONALD, DAVID, (ed.) *The Encyclopaedia of Mammals.* Unwin Hyman Ltd, London, 1989.

MAYNARD SMITH, J., *The Evolution of Sex.* Cambridge University Press, Cambridge, 1978.

MCLAREN, ANNE, (ed.) *Advances in Reproductive Physiology.* Academic Press, London, 1966.

MORRIS, DESMOND, *The Human Animal.* BBC Worldwide Ltd, London, 1994.

NULAND, SHERWIN B., *How We Die.* Vintage, London, 1996.

ROSS, JANET S., & WILSON, KATHLEEN J.W., *Foundations of Anatomy and Physiology.* Churchill Livingstone, Edinburgh, 1964.

SINCLAIR, DAVID, *Human Growth After Birth.* Oxford University Press, Oxford, 1991.

SMITH, ANTHONY, *The Body.* George Allen & Unwin Publishers Ltd, London, 1985.

SMITH, ANTHONY, *The Mind.* Hodder & Stoughton, London, 1984.

SMITH, ANTHONY, *Sex, Genes and All That.* Macmillan Publishers Ltd, London, 1997.

SOKOLOV, IVAN, & HUTTON, DEBORAH, *The Parents Book.* Thorsons, London, 1988.

STACEY, MEG, (ed.) *Changing Human Reproduction.* Sage Publications Ltd, London, 1992.

VEATCH, ROBERT, M., *Death, Dying and the Biological Revolution.* Yale University Press, New Haven, US, 1976.

WARNOCK, MARY, *A Question of Life.* Blackwell Publishers, Oxford, 1992.

WATSON, E.H., & LOWREY, G.H., *Growth and Development of Children.* Year Book Medical Publishers, Chicago, US, 1962.

WATSON, JAMES D., *The Double Helix.* Penguin Books Ltd, Middlesex, 1962.

WILSON, KATHLEEN J.W. Churchill Livingstone, Edinburgh, 1964.

WINGATE, PETER, *The Penguin Medical Encyclopaedia.* Penguin Books Ltd, Middlesex, 1976.

WOLPERT, LEWIS, *The Triumph of the Embryo.* Oxford University Press, Oxford, 1993.

WORLD BANK AND WORLD HEALTH ORGANISATION, *Global Burden of Disease Study.* 1997.

YOUNG, J.Z., *Programs of the Brain.* Oxford University Press, Oxford, 1978.

Picture credits

© BBC: 16–17 David Barlow; 24 Richard Dale; 54 Jane and Richard Jeffers; 86 Peter Georgi; 120 Andrew Thompson; 132 (above); 154–5 (inset) John Green; 156 Chris Spencer; 167 & 171 David Barlow; 186 Chris Spencer; 194 (left) David Barlow.

© Bubbles: 66 Ian West; 80 J. Farrow; 81 Frans Rombout; 93 (above) & 100 (inset) Denise Hager.

© Collections: 82 Fiona Pragoff.

Getty Images: 31, 35 (inset), 39, 43 (left), 52–3 (inset), 59, 64, 74–5, 93 (below), 118–9 (inset), 122, 126, 144–5, 169 (right), 176 (below), 178 (inset), 182, 204–205.

© Zena Holloway: 64–5 (model from Little Dippers).

Image Bank: 170.

Rex Interstock Ltd: 112 (inset).

© Sally & Richard Greenhill: 61 (above).

Science Photo Library: 6–7, 8–9, 12–13, 20–21, 28–9, 32–3, 37, 41, 42, 43 (right), 46, 47 (left), 49, 51, 52–3, 61 (below), 70–1, 84–5, 90–1, 94, 99, 100–101, 104–5, 108–9, 112–3, 118–9, 127, 132 (below), 134-5, 138-9, 142, 143, 146-7, 148-9, 154-5, 159, 164-5, 166, 169 (left), 172-73, 176 (above), 177, 178, 184–85 (inset), 188, 189, 194–95, 198–99, 205.

Courtesy Vertec Scientific: 115.

Wellcome Trust Photographic Library: 22–23 (inset), 34–35, 47 (right), 56, 184–85.

Zefa: 19, 22–23.

Index

Page numbers in *italics* refer to the illustrations

abdominal pregnancy 35
acne 94, 110
Adam's apple 102
adolescence 93–98
adrenalin, in newborns 55, 57
afterbirth 37
ageing 155–183
  attitudes to 182–183
  effects on body 106, 168
  effects on bones 176–177, *178–179*
  effects on brain 160–165
  effects on hair 166–167
  effects on sight and hearing 168–169, *172–173*, 174
  effects on skin 170–171
  hormonal changes in men 181–182
  menopause 175, 180–181
  theories of 157–160
algae 14
allergy 143
Alzheimer's disease 164–165
amniocentesis 45
amniotic cavity 36, *37*
anaerobic respiration 207
anaphylactic shock 143
aneurysms, surgical removal 201
angina 205
angiogram *185*
antibodies 60, 142
antigens 142
antioxidants 160
aorta 126
apocrine glands 100
apoptosis 194, *195*
armpit hair 103, 110
arteries 126, *185*
arterioles 126, *127*
arteriosclerosis 204
arthritis 176–177
atherosclerosis *184–185*, 204
ATP 197
auto-immune diseases 143

babbling 76, 129–130
babies and children 52–79
  brain development 69, 72
  learning language 76–79
  learning to walk 74–75
  manipulation of adults 67–69
  mental development 79–83
  mortality 53–55
  newborn 45, 55, 56, 57–61, 63–68
  teeth 70–71
  toilet training 73, 76
  see also birth
bacteria, in gut 60
balance organ
  see vestibular apparatus
baldness *166*, 167
beard (facial) hair 100, *101*, 103
bipedalism *53*, 74, 123
birdsong 129
birth 24, 54, 55–57, 60
  baby's size at 62–63
blackheads 94
blood 126
blood pressure, in babies and adults 58
blood vessels 126, *127*
body height 111, 114
body odours 100
body temperature, control in babies 58, *59*
body weight
  at birth 62–63
  at onset of puberty 87–88
  gain in pregnancy 51
bone cell *112–113*
bones *see* skeleton
brain 10, 18, *119*, 121–125, 128
  changes with ageing 160–165, 174
  cryosurgery 200
  damage to 152
  demand for oxygen 197
  development 38, 60–61, 62, 69, 72, 194
  effects of oxygen starvation 209
  effects of stroke 205
  how it functions 137, 139
  left–right asymmetry 134–136
  and mental abilities 152–153
  and 'mindreading' 80, 83
  role in automatic actions 141
  role in ejaculation 33
  role in puberty onset 86, 87
  role in sexual arousal 97
  role in vision 150
  signals and connections 136–137
  waves 136
brain cells (neurones) *6*
  connecting pathways 124–125, *138–139*
  development 69
  effects of ageing 161
  signalling by 137
brain death 202–203
brain stem 124, 202
Braxton Hicks contractions 47
breast
  changes with age 106
  changes at puberty 106, 107
  changes with menstrual cycle 106
  changes in pregnancy 48, 50, *51*, 106
  male 103, 107
breathing
  believed essence of life 200–201
  establishment after birth 56, 57

cancer 156, 158, 188–189
  cause of death 186, 191
  of lung 188
  of skin 171, 188
capillaries 126
cataract 169
cave-dwellers 17–18
cell death *see* apoptosis; necrosis
cell division *159*
cellular respiration 197
cerebellum *119*, 141
cerebral hemispheres 10, *119*, 124, 134–136
  visual cortex 145
cerebrovascular disease (stroke) 205
cerebrum 123
Cheyne-Stokes respiration 206
childhood *see* babies and children
cholesterol 204
chromosomes 11, *12–13*, 32
  see also sex chromosomes

circulatory system 126, *127*
  changes at birth 57
collagen 170
colour-blindness 150–151
comedones 94
conception
  see fertilisation
cone cells *147*, *148–149*
consciousness 153
corona radiata 27
coronary arteries *185*
coronary heart disease 204–205
corpus luteum 108
crawling 74–75
crying 67
cryosurgery 200

death 185–209
  body changes following 207–208
  causes of 191–193, 204–205
  of cells 194–195
  changing patterns 187, 190–193
  diagnosis 200–203
  final moments 206
  and oxygen deprivation 196–200
  as part of life 193, 196
delinquency, in adolescence 95
dementia 163–165
diet
  and atherosclerosis 204
  and cancer 160, 189
diving reflex 64, *65*
DNA (deoxyribonucleic acid) 11, *154*
  damage to 159, 160, 189
double mating 40
dreaming 68

ear
  evolution 15, 17
  organ of Corti 74, *172–173*, 174
  vestibular apparatus 52, 74
ear bones 15–17, 174
'ear dust' 53, 74
eating *see* nutrition
eccrine glands 100
ectopic pregnancy 35
egg *22–23*, 32
  see also fertilisation
ejaculation 33, 103
elastin 170
embryo, development 17, *34*, 35–38, 42–43
energy, body's supplies 197
epididymis 33, *99*
erection 98, 102
erythrocytes
  see red blood cells
evolution (human) 11, 14–15, 156–157
  bipedalism 121, 123
  brain 10, 121, 125, 128
  ear bones 15, 17
  menopause 181
  role of death in 196
  timing of sexual maturity 89, 92
eyes 144–145, *146–147*, *148–149*
  changes with ageing 169

fat, distribution in body *115*, 116
fertilisaton 26–35
foetus *23*, 43
  development 44–48, 194
  movements 25, 41, 44
  sleeping 68
follicle-stimulating hormone (FSH) 86, 87, 108
fraternal (non-identical) twins 39, 40, 114

freckles 170
free radicals 160
'frobbing' 68
FSH (follicle-stimulating hormone) 86, 87, 108

gender ambiguity 8–9
gender determination
  see sex chromosomes
gender differences
  behaviour 89
  birth and death rates 190–191
  brain 89, 123
  colour-blindness 151
  fat distribution 115, 116
  osteoporosis incidence 176
  puberty 85, 86–89, 98, 102–103, 106–110
  skeletal structure 114
  teeth development 71
  see also sex chromosomes
gene combinations 32
genetics 11
  no role in ageing 157
gestures 128–129
gonadotrophin releasing hormones 86
grasping reflex 64
growth hormone 111
growth spurt 103, 110, 111–116

haemoglobin 126
hair 100, *101*, 103, 110
  greying 167
  loss 166–167
hand 132
  development *61*, 132, 194
  evolution 123
handedness 131, 133
HCG (human chorionic gonadotropin) 108

head *119*
  baby's 55, 60, 62–63
hearing
  baby's 63
  in foetus 44
  loss with age 174
heart *20–21*, 126, 204
  changes at birth 57
  development 38
heart disease 204–205
  detection *185*
heart rate, in babies and adults 58
hip replacement *176*
hormone replacement therapy 176, 180
hormones
  changes with ageing 175, 180, 181
  see also individual hormones
'hot flushes' 175
hypophysis
  see pituitary gland
hypothalamus 86, 87

identical twins 39, 40
  body height 114
  handedness 133
  movements in uterus 41
  understanding mirror images 80
imaging techniques *8–9*
  see also magnetic resonance imaging
immune system 142–143
  baby's 60
  damage by ultraviolet light 171
implantation 30, *34*, 35
infant mortality 53–55
infectious diseases 191
infertility 33
interferons 142
ischaemic heart disease 204–205

lactation 106
male (reported) 107
language
76–79, 128–131
larynx
baby's 129
changes at puberty
102, 110
learning 66, 73, 76, 120
language 76–79
not to think 141
to walk 74–75
left-handedness 131, 133
leukocytes
see white blood cells
LH (luteinising hormone)
86, 87, 108
life expectancy 187, 190
lifespan 157–158, 190
liver spots *155*, *170*, 171
lung
cancer of 188
changes at birth 57
evolution 15
luteinising hormone (LH)
86, 87, 108

macrophages 142, *143*
magnetic resonance
imaging
8–9, 80, *119*,
134–135
male pattern baldness
166, 167
mammary line 106
melanin 151, *155*, *170*
memory
'experts' 152
loss with age 161
memory cells 142
menarche 86, 87, 110
menopause 175, 180–181
menstruation
106, 108–109, 116
cessation with
pregnancy 26
start of *see* menarche
synchronisation 92
metastasis 188
mind 79–83, 117,
152–153, 161
'mindreading' 80, 83
mirror images,
understanding
79–80
miscarriage 36, 196
mitochondria
197, *198–199*
multiple births 40
mutations 158, 189

'Near Death Experiences'
208, 209
necrosis 194
nerve cells *see* neurones
neurones 118
see also brain cells
neurotransmitters 137
nutrition
adolescents' 116
babies' 68, 116
and onset of puberty
87–88
see also diet

oestrogen
crystal form *90–91*
production at puberty
87, 106
production fall at
menopause 175
role in acne 94
role in menstrual cycle
108, 175
use in HRT 180
organ of Corti
74, *172–173*, 174
osteoarthritis 176–177
osteoporosis
176, *178–179*, 180
Otzi ('ice man') 208
'out of body experiences'
209
ovary 88
changes at puberty
107
role in menstrual cycle
108
ovulation 26, 27, 92,
108, *109*
oxygen 196–197, 200
effects of starvation
209

pacemaker 126
pain, foetal perception 45
'Peat Bog' 208
pelvis, changes with
puberty 110, 114
penis 33, 98, 102
pentadactyl limb 15, 123
persistent vegetative state
203
pheromones 92
pituitary gland *84–85*,
86, 87, 108, 111
placenta 36–37
polygyny 182
polythelia 106
pregnancy 23–26,
48–51, 106
after menopause 175
duration 26, 42
food craving/aversions
in 48, 50
sickness in 48, 50
teenage 97
weight gain in 51
pregnancy tests 25–26
premature babies 58, *59*
pre-menstrual syndrome
108
progesterone 87, 106,
108, 175, 180
programmed cell death
194, *195*
prolactin 87
prostaglandins 27
psychiatric disorders
(in adolescence) 95
puberty 84–117
in females 87, 106–110
in males 86–87,
98, 102–103
pubic hair 100, 103, 110

'quickening' 25, 44
'rate of living' theory
(ageing)
157–158
red blood cells 126
reflexes, baby's
64–65, 67
REM sleep 68
respiration 197
anaerobic 207
retina *147*, *148–149*, 150
rigor mortis 207
rituals of puberty
96
rod cells *147*, *148–149*
rooting reflex 64

'Sally-Anne' test 80
seminiferous tubules
32, 98, *99*
senility 163
sex chromosomes
12–13, 30, 85, 88
see also X chromosome
sexual intercourse,
in adolescence
96, 97–98
Siamese twins 39
skeleton
bone cell *112–113*
changes with ageing
176–177
development
19, 47, 61, 111, *132*
evolution 123
skin
body's defence
142
cancer of 171, 188
change at puberty
103
changes with ageing
155, 170–171
skull, baby's 60
sleep 68
smell, in foetus 45
smiling 67
smoking, cancer risk
from 188
social skills 79–83

sperm *22–23*, 32–33, 98, 103, 181
  see also fertilisation
spots *see* acne
stress hormones, in foetus 45
stroke 205
suicide 191
  in adolescence 95
swaddling *59*
sweat glands 100, 103, *105*
sweating 100, *104*, 110
synapses 69, 125

taste, in foetus 45
teeth 70–71
testis, testicle 98, *99*
  changes at puberty 102
  descent 98
  development 88
  sperm production in 32–33
testosterone
  crystal form *91*
  and larynx enlargement 102
  production at puberty 86–87
  role in acne 94
  role in sexual differentiation 88, 89
toilet training 73
tools, prehistoric stone 130, 132
twins 39–41
  body height 114
  handedness 133
  menarche 110
  understanding mirror images 80

ultrasound scanning 24, *36*
ultraviolet radiation, effect on skin 170, 171
uterus 30
  changes at puberty 107
  changes with menstrual cycle 108

vaccination 142
vagina, changes at puberty 107
'vanishing twin' syndrome 40
varicose veins 126
vas deferens 33, *99*
veins 126
vena cava 126
vestibular apparatus *52*, 74
vision 144–151
  baby's 63, 66
  changes with ageing 168–169
  development 69, 72
visual cortex 145
vocal cords 102, 129
voice, change at puberty 102, 103, 110

walking, learning *53*, 74–75
weightlessness 120
'wet dreams' 103
white blood cells (leukocytes) 7, 126, 142, *143*, *189*
  apoptosis *194–195*
'witches' milk' 106

X chromosome *12–13*, 30, 88, 151, 167

Y chromosome *12–13*, 30, 88